AF679081

CORROSION INHIBITORS IN CONSERVATION

Occasional Papers Number 4 1985

The Proceedings of a conference held by UKIC in association with the Museum of London

Editor: Suzanne Keene

General Editor: Katharine Starling

Contents

ISBN 0 9504155 4 5

Preface

These papers were given at a conference held jointly with the Museum of London in February 1984. Help given by the following institutions is gratefully acknowledged: the British Academy and the British Council, for grants for overseas speakers; and the Museum of London for providing all the necessary staff time and premises for the meeting. Helen Ganiaris, Rose Johnson and Kate Starling undertook most of the work, with great efficiency.

Suzanne Keene
September 1984

HOW CORROSION INHIBITORS WORK

Brian S. Skerry

CORROSION PROCESSES

Corrosion can be considered as the spontaneous reversion of a metal to its natural valency state by reaction with its environment. Whereas thermodynamic factors determine whether or not a metal can corrode, the rate of such a reaction is determined by kinetic factors. Although the tarnishing of some metals such as silver occurs in the absence of an electrolyte, this effect should more accurately be described as oxidation rather than corrosion. Corrosion processes are generally considered to be electrochemical in nature. The necessary liquid phase electrolyte may occur as a thin moisture layer which adsorbs onto the metal surface from the atmosphere, or as moisture in soils, or indeed, the corroding metal may have been placed in natural waters, or in aqueous solutions of cleaning agents. When a metal corrodes in an electrolyte, two coupled reactions are required to proceed at the metal surface: an anodic reaction consisting of the transfer of metal ions from the lattice to the solution with consequent production of electrons, and a cathodic reaction consisting of the assimilation by some agent of the electrons liberated by the anodic reaction. A corrosion process is typically represented by two half reactions;

$M \rightarrow M^{n+} + ne^-$

anodic (oxidation) reaction

$Ox + ne^- \rightarrow Red$

cathodic (reduction) reaction

Certain chemical compounds or mixtures of compounds are known to reduce the rate at which corrosion processes occur, or to suppress them completely, when introduced in small concentrations (typically of the order of 50-500 ppm) to the 'corrosive' environment. The ability of such 'inhibitors' to reduce rates of corrosion is generally considered in terms of efficiency, calculated as:

$$\text{Inhibitor efficiency (\%)} = \frac{CR - CR_i}{CR} \times 100$$

where CR = corrosion rate without inhibitor
CR_i = corrosion rate with inhibitor

The value can be high, over 99% given favourable circumstances.

There is some difficulty in describing the precise chemical nature of many inhibitor formulations that are in general use, since most corrosion inhibitor manufacturers do not divulge the compositions of proprietary inhibitors. Considerable research, however, has been carried out to determine which compounds or functional groups perform the protective action in complex compositions, and several reviewers have sought to rationalise and correlate mechanisms of corrosion inhibition. A number of books have been devoted entirely to the subject of corrosion inhibition.[1-3]

In order to establish the mechanisms by which inhibitors function, it is necessary to consider the nature of corrosion processes in terms of the chemical environment, as well as both thermodynamics and kinetics.

THE CHEMICAL ENVIRONMENT

Metallic corrosion in aqueous solutions of pH between 5 and 9 is considered to be taking place in neutral conditions. Corrosion in such neutral media is very common and is widely observed, for example, in rain water, river water, moist soil and sea water. The corrosion products formed are usually sparingly soluble, for example iron hydroxides (rust), basic zinc, lead or copper carbonates and aluminium hydroxide. The resultant partial screening tends to protect the metals from further corrosion. However, such protective effects are not great and hygroscopic corrosion products may accelerate corrosion.

Aqueous solutions of acids are among the most corrosive liquids known, and metals can be destroyed at a very high rate in acidic conditions, especially when soluble products are formed.

In alkaline conditions, amphoteric metals, aluminium, zinc and copper, are attacked with particular vigour. However, the alkaline corrosion of iron is comparatively slow and, in some circumstances, alkalinity can induce inhibitive secondary effects. For example, if calcium carbonate is present in solution, and the pH becomes high enough, a layer of calcium carbonate may be deposited, which physically segregates the metal from the corrosive environment.

THERMODYNAMICS OF CORROSION PROCESSES

When metals are reduced from their ores, a natural reaction is reversed. Hence metallic corrosion is a thermodynamically favourable process since the metal is merely reverting to its inherently most stable form. Immersed in, or covered by a layer of, electrolyte a metal acts as an electronic conductor or electrode.

For a given electrode reaction such as

$Ox + ne^- \rightarrow red$

the electrode potential, E, can be shown to depend on the concentrations of reduced and oxidised species:

$$E = E^\circ + \frac{2.3\ RT}{nF} \log_{10} \frac{[ox]}{[red]} \quad \text{(the Nernst Equation)}$$

E° is the standard electrode potential (ie for a pure metal with metal ions at unit activity in solution), R is the gas constant, T the absolute temperature, n the change of ionic charge in the electrode process and F represents Faraday's constant.

For many reaction processes the Nernst equation indicates pH dependence. For example, in the case of

$Fe_3O_4 + 2H_2O + 2e^- = 3HFeO_2^- + H^+$

the Nernst equation becomes (at 300 K),

$E = E^\circ + 0.0298pH - 0.089 \log [HFeO_2^-]$

(since $pH = -\log_{10} [H^+]$)

Using the Nernst equation and measurable solubility data for various metal compounds, diagrams of potential versus pH have been drawn up for many metals[4], including iron and copper (Fig. 1). These potential-pH diagrams are very useful in corrosion studies as they give rapid indications as to whether reactions are thermodynamically possible.

Consider, for example, an iron specimen immersed in an oxygen free solution at pH 7 (Fig. 1a). The iron corrosion potential would occur between the horizontal line (representing the equilibrium between the metal and its ions in solution) and the sloping, dashed line (corresponding to the hydrogen electrode potential). Three possible actions (denoted by arrows 1, 2 and 3) could be taken to remove the specimen from the corrosive region:

1. The electrode potential may be lowered (by making the iron the cathode in a galvanic cell) so the specimen attains corrosion immunity.

2. The electrode potential may be raised (by adding

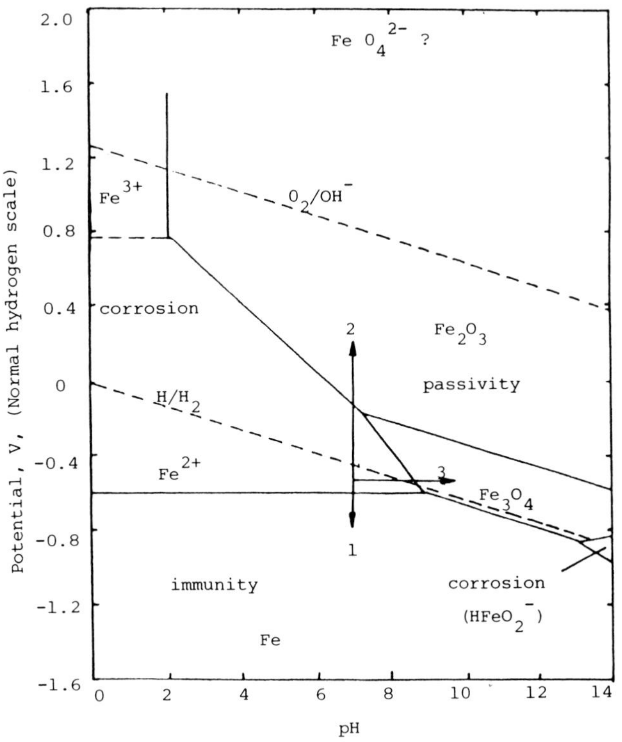

Fig 1a Potential - pH diagram for iron (10^{-6} gram atoms . φ^{-1}, 25°C)

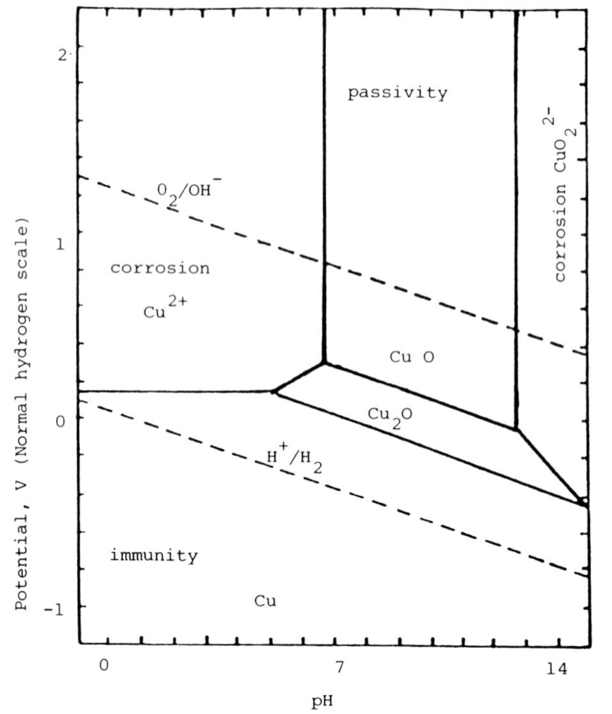

Fig 1b Potential - pH diagram for copper (10^{-6} gram atoms . φ^{-1}, 25°C)

oxidising agents) which makes the specimen passive.

3. The pH may be raised to provide a degree of specimen passivity.

In the case of copper (Fig. 1b) in deaerated solutions of non-oxidising acids, the potential of the hydrogen electrode falls within the region of immunity for all pH values, and copper will not be attacked. In aerated or oxidising solutions, however, copper is attacked forming Cu^{2+} in acid or CuO_2^{2-} in alkaline conditions.

Although potential-pH diagrams are useful in terms of what theoretically can or cannot happen, they have inherent limitations which should always be considered. The diagrams presuppose equilibria between the metal and its ions. In practical corrosion cases, however, conditions are rarely at equilibrium. Moreover, the diagrams take no account of other ions which may be present (eg sulphides in soil), local environment effects (eg localised pH) and, most importantly, they do not predict the rates at which reactions proceed. Indeed, reaction kinetics are of fundamental significance when corrosion processes are considered.

KINETICS OF CORROSION PROCESSES

When a metal becomes covered by or immersed in an electrolyte, charges of opposite sign normally accumulate near the electrode/electrolyte interface. This occurs because as the metal dissolves to form aquo-ions (M^{n+}), there is a separation of electric charge, the metal having a net negative charge and the aqueous solution adjacent to it a positive one. Such a system is called an electrode. The build-up of charges gives rise to a potential difference between the electrode and the electrolyte. Although several models have been proposed to illustrate this region of charge separation or electrical double layer, which are discussed in the literature[5,6] it is generally accepted that the double layer comprises a strongly bound inner plane of ions adjacent to the electrode surface and a much more weakly bonded outer, diffuse layer of ions.

At equilibrium, the rate of electron transfer across a metal electrode/electrolyte interface in one direction is balanced exáctly by an equal rate of electron transfer in the opposite direction so that current flows can be represented by

$$\overrightarrow{i} = \overleftarrow{i} = i_o$$

where i_o is the exchange current density.

Also, the equilibrium potential is determined by this condition. However, when a metal corrodes this equilibrium condition is no longer valid, the net current density being represented by

$$i = \overrightarrow{i} - \overleftarrow{i}$$

where the electrode potential difference departs from the equilibrium value $\Delta\phi_e$ to a value $\Delta\phi$. The difference between the new value and the equilibrium value is referred to as the over-potential, η. Thus,

$$\Delta\phi - \Delta\phi_e = \eta$$

In the case of either an anodic over-potential, η_a, or a cathodic over-potential η_c, the net anodic and cathodic currents are represented by the following equations (known as the Butler-Volmer equations) respectively[7]:

$$i_a = \overrightarrow{i} - \overleftarrow{i}$$

$$= i_o \ \exp\left[\frac{nF\beta\eta_a}{RT}\right] - \exp\left[\frac{-nF\ (1-\beta)\eta_a}{RT}\right]$$

$$i_c = \overleftarrow{i} - \overrightarrow{i}$$

$$= i_o \quad \exp\left[\frac{-nF\ (1-\beta)\eta_c}{RT}\right] - \exp\left[\frac{nF\beta\eta_c}{RT}\right]$$

where β (and $1 - \beta$) is the fraction of the total distance across the electrical double layer to where the peak of the activation energy hump occurs. Other symbols are as defined earlier.

Now, for high field approximations where $\overrightarrow{i} >> \overleftarrow{i}$:

$$i_a \approx i_o \exp\left[\frac{nF\beta\eta_a}{RT}\right]$$

and where $\overrightarrow{i} << \overleftarrow{i}$

$$i_c \approx i_o \exp\left[\frac{-nF\ (1-\beta)\eta_c}{RT}\right]$$

By rearranging, taking logs, and allowing

$b_a = 2.3RT/nF\beta$ and $b_c = -2.3RT/nF(1-\beta)$,

$\eta_a = b_a \log i_a/i_o$, and

$\eta_c = b_c \log i_c/i_o$

These equations are known as the Tafel equations after Tafel, who discovered the relationship in 1905.[8] The implications of the Tafel equations are that for values of over-potential (in practice) $\gtrsim 30^{m}V$, there is a linear relationship between over-potential and the logarithm of the current (Fig. 2).

Thus far, only half-cell reactions have been considered. However, in reality, the spontaneous corrosion process of a metal in an electrolyte is a combined effect of two half cell reactions. In the case of iron corrosion in hydrochloric acid, for example:

$Fe \rightarrow Fe^{2+} + 2e$ (anodic half cell reaction)

$2H^{+} + 2e \rightarrow H_2$ (cathodic half cell reaction)

each half cell reaction has its own respective equilibrium potential. Since the hydrogen evolution (cathodic) reaction takes place on the iron substrate, the resulting corrosion (mixed) potential will be somewhere between the equilibrium potentials of the two half cells. For corrosion to proceed, the anodic and cathodic reaction rates must be equal and this situation can be illustrated by a diagram of potential against current (Fig. 3).

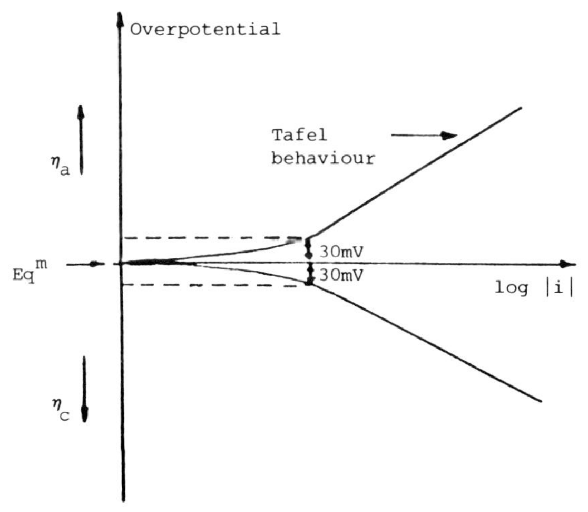

Fig 2 Relationship between overpotential and net or applied current density.

These potential/current diagrams were first used by U.R. Evans and are generally referred to as Evans diagrams.[9] In such a diagram the equilibrium potentials for the cathodic and anodic reactions are shown. The potential E_{corr} represents the mixed (or combined) potential and the rate of corrosion reaction is represented by the corrosion current (i_{corr}) which can be related directly to metal dissolution by Faraday's Law. Evans diagrams, although theoretical, are based on the Tafel slopes for the system under investigation, deduced experimentally by means of polarisation curves. The advantage of Evans diagrams is that they can be used to rationalise the electrochemical effects of additional chemicals introduced to a corroding system. The changes in polarisation curves caused by such additions affect Tafel slopes. Any consequent effects on either E_{corr} or i_{corr} can be established by constructing Evans diagrams for the system. A reduction in i_{corr} suggests a lowered corrosion rate, while the nature of the change in E_{corr} can provide the information as to the induced mechanisms.

CORROSION INHIBITORS

CLASSIFICATION OF INHIBITORS

There are many types of chemicals which inhibit corrosion processes under different circumstances and some method of classification of inhibitors is required.

Methods used to classify inhibitors are somewhat arbitrary and are not entirely satisfactory. Several classes have been suggested, such as anodic, cathodic, neutralising, passivating, precipitation, vapour phase or adsorbants. Some inhibitors could, however, be classified in two or more of these categories.

Indeed, Evans[10] considers that all inhibitors could be termed anodic, cathodic or mixed in view of the fact that the first type predominantly retards the anodic reaction, the second type retards the cathodic reaction, while the third type retards both reactions simultaneously. Thus, there is no standard method of classification which is universally accepted.

For the purposes of the present discussion, inhibitors may be conveniently classified according to three variables:

i. the way in which they act, ie by affecting the anodic or cathodic reaction;

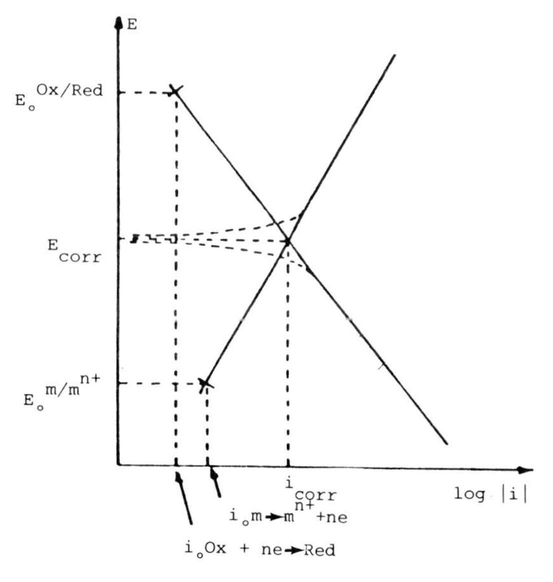

Fig 3 Potential - current diagram illustrating the behaviour of a corroding metal

ii. their chemical composition, ie inorganic or organic compounds;

iii. the type of environment in which they are effective, ie acid, neutral or alkaline.

With regard to mechanisms of inhibition, information can be obtained from a study of Evans diagrams for specific systems selected from combinations of these categories. If, for example, anodic or cathodic potentials are observed to be altered, this indicates interference with the thermodynamics of the system, whereas if the slope of either curve is altered, this indicates changes in the kinetics of the system. It should be noted, however, that the simultaneous action of two or more mechanisms is common, even with single component inhibitor systems.

CATHODIC INHIBITION

In near neutral solutions, the preferred cathodic reaction is commonly the reduction of dissolved oxygen:

$O_2 + 2H_2O + 4e^- \rightarrow 4OH^-$

In deaerated solutions, however, the only available reaction, in the absence of easily reducible species, is H_2 evolution:

$2H^+ + 2e^- \rightarrow H_2$

The change from one of these reactions to the other results in a change in the cathodic reaction potential. Thus, deaeration by itself could be considered as an inhibition process, and oxygen scavenging agents such as hydrazine or sodium sulphite are correctly termed 'inhibitors' in this sense. The effect of reducing the cathodic potential can be illustrated by an Evans diagram (Fig. 4). The reduction in corrosion current, $i_{corr} \rightarrow i_c'$, indicates a lower corrosion rate. Such an effect represents a change in the system's thermodynamics.

In practice, however, the effects of deoxygenation are not confined to the thermodynamics of the cathode reaction. There is also an effect on the polarisation of the cathode, causing the slope of the cathodic curve to be changed. This indicates an effect on the system's kinetics.

In the case of iron corroding in acid, the dissolution of iron is accompanied by the evolution of hydrogen. This takes place at cathodic sites in two stages:

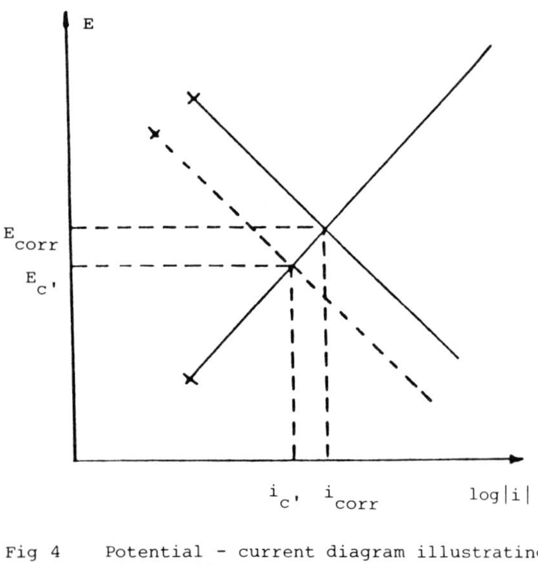

Fig 4 Potential - current diagram illustrating a change of cathodic reactions in the presence of an inhibitor

1. $H^+ + e^- \rightarrow H$

2. $2H \rightarrow H_2$

Some substances 'poison' the second reaction and the overall corrosion becomes rate limited by the accumulation of monatomic hydrogen at the metal surface. This kinetic effect markedly alters the slope of the cathodic curve (Fig. 5). The resulting decrease in the corrosion current ($i_{corr} \rightarrow i_c'$) indicates less severe corrosion, whilst the simultaneous decrease in the corrosion potential ($E_{corr} \rightarrow E_c'$) is indicative of retardation of the cathodic reaction.

Typical examples of this inhibitive action are provided by the restrainers that are sometimes used to minimise dissolution of metal in pickling acids or in de-rusting acids, for example, antimony trichloride or stannous chloride in hydrochloric acid, and sodium chloride in sulphuric acid.

In near-neutral aerated solutions, true cathodic inhibitors form surface films of insoluble salts. For example, salts of metals such as zinc, magnesium, manganese and nickel form insoluble hydroxides at cathode sites. Here, localised alkalinity is caused by an increase in hydroxyl ion concentration following the oxygen reduction reaction. Since these insoluble precipitates are poor electronic conductors, oxygen reduction does not occur on their surfaces. The precipitates also restrict oxygen diffusion to the metal surface. Since the cathode has become blocked, the cathodic site moves during the next $Fe \rightarrow Fe^{2+}$ transition, progressively covering the entire surface with the insoluble precipitate. These inhibitors are termed 'safe' inhibitors, as even if the precipitate film becomes flawed, increasing alkalinity at the flaw will tend to produce self-healing of the film. In conditions where the concentration is too low to be effective, a 'safe' inhibitor will encourage uniform corrosion over the surface, rather than localised corrosion, at an overall rate no greater than in the absence of the inhibitor.

ANODIC INHIBITION

A considerable number of chemical compounds act as inhibitors for the corrosion of metals in near-neutral solutions by stabilising naturally occurring oxide films, forming 'passivating' films. These compounds are generally termed 'anodic' inhibitors as they are considered, principally, to affect anodic processes. In the presence of these inhibitors, the oxide surface film on the metal tends

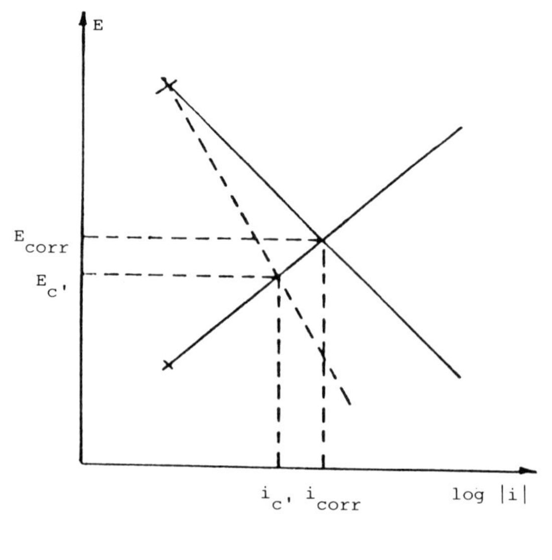

Fig 5 Potential - current diagram illustrating inhibitor induced cathodic polarisation

to be thicker and more continuous with the metal at a higher oxidation state. The surface is said to be passivated because the potential of the metal/metal oxide surface is raised. This effect is illustrated by the potential-pH, or Pourbaix, diagram for iron (Fig. 1a), whereby the free corrosion potential can be shifted into the passive zone. It can also be seen from the Pourbaix diagram that such inhibitors act only at moderate pH, as oxide films are not stable at very high or very low pH.

Inhibitors affecting the anodic process fall into two distinct categories, commonly referred to as oxidising and non-oxidising. Chromates [CrO_4^{2-}] and nitrites [NO_2^-], for example, are oxidising anions and can inhibit efficiently even in the absence of oxygen. Phosphates [PO_4^{3-}] and silicates [SiO_3^{2-}] and anions of certain organic acids, for example benzoate and azelate, act efficiently only in the presence of oxygen and are classified as non-oxidising inhibitors.

The anodic polarisation diagram of a steel in a water containing oxidising inhibitors provides a possible explanation for the inhibitive behaviour of oxidising anions (Fig. 6). Without inhibitors, the steel corrodes in the active state at a rate corresponding to point A. In the presence of the oxidising inhibitor, however, providing it is in sufficient concentration, there is a transition in corrosion potential from E_{corr} to $E_{c'}$, allowing passivation to occur. This follows from mixed-potential theory, as the steady state is determined by the intersection of total oxidation and reduction rates. Moreover, the corrosion current is reduced from i_{corr} to $i_{c'}$.

Although these oxidising inhibitors can act in the absence of oxygen, their disadvantage is that if present in insufficient concentrations they passivate only parts of the surface: the remainder corrodes at a greater rate than in the absence of the inhibitor, often resulting in severe pitting. These inhibitors are sometimes known as 'dangerous' inhibitors.

Non-oxidising inhibitors require relatively higher concentrations than oxidising inhibitors. They also require the presence of dissolved oxygen to act efficiently. As with oxidising inhibitors, the film is kept in good repair by oxidation of the metal (by oxygen) and by local precipitation of salts of the inhibitive ion. In effect, non-oxidising anions modify the anodic kinetics by reacting with metal ions to form species which are more protective than the salts which would otherwise form.

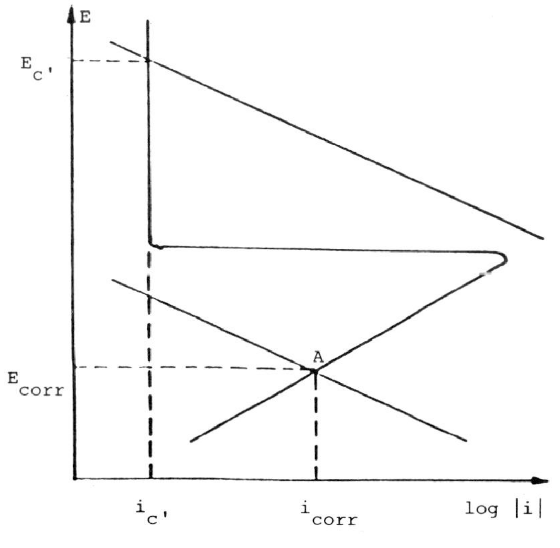

Fig 6 Anodic polarisation diagram illustrating inhibitor induced passivation

An interpretation using potential-current diagrams is also possible, illustrated by a change in the slope of the anodic kinetics curve (Fig. 7). The accumulation of primary corrosion oxides causes polarisation of anodic sites. If these oxides are stabilised by insoluble secondary products, increased polarisation and consequent inhibition is the result.

LIMITATIONS OF USING POTENTIAL-CURRENT DIAGRAMS FOR EXPLAINING INHIBITOR MECHANISMS

Schematic potential-current diagrams show some serious limitations if they are used to construct a generalised model for explaining inhibitor mechanisms. In theory, all oxidising anions should have some inhibitor efficiency, but in practice they do not. Moreover, the model does not suggest that the oxidising inhibitor (or its product) could be found in the passive film, but in practice they often are.

In the case of non-oxidising inhibitors, the model does not explain the need for non-oxidising anions, since the presence of sufficient oxygen should result in a passive protective film. In addition, the simultaneous use of two inhibitors may synergistically rather than additively improve the inhibitor mixture efficiency and may eliminate the risk of pitting at low inhibitor concentrations. (Mixtures of oxidising anions with non-oxidising, precipitating anions, such as chromate with orthophosphate, are particularly effective even in sea water.)

It is likely that the inhibitor ions may have some effects at the interface not considered in potential-current diagrams. For example, they may act as buffers controlling the pH level at the interface, or mechanisms of surface adsorption may be of overriding significance in some circumstances. Consequently, other models which have been developed to explain inhibitor mechanisms require consideration.

1. Inhibitors as Buffering Agents

It has been suggested that the action of the inhibitor could be to control (buffer) the pH at the metal/electrolyte interface, preventing acidity. This would stabilise the protective oxide film. For example, Hoar[11] has shown that for tin in solution, at an oxide film flaw, tin becomes oxidised:

$$Sn \rightarrow Sn^{4+} + 4e^-$$

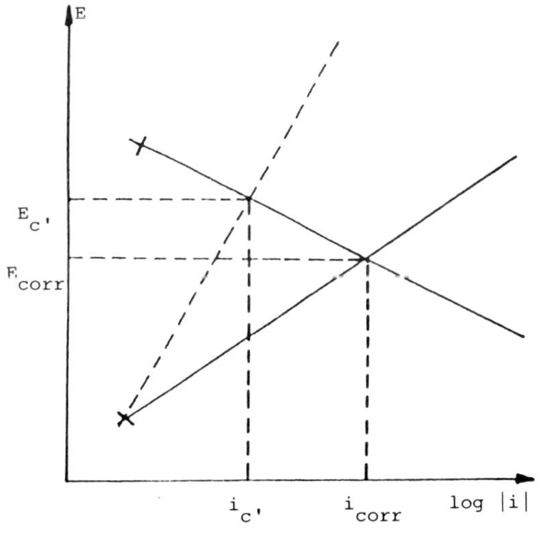

Fig 7 Potential - current diagram illustrating induced anodic polarisation

which, following further reaction with water, gives tin hydroxide:

$Sn^{4+} + 4H_2O \rightarrow Sn(OH)_4 + 4H^+$

thus increasing the concentration of hydrogen ions. This build up of acid normally undermines the tin oxide film, leading to corrosion. In the presence of a buffering inhibitor, the depassivation reaction at the anodic site may be stopped by the inhibitor preventing a build up of acidity. Although the buffering concept as developed by Hoar[12] is helpful, it is limited in its usefulness since not all inhibitors are buffers. For example, some salts of fairly strong acids such as substituted benzoates have no buffering action but are powerful inhibitors. Moreover, not all buffers are inhibitors. For example, some salts of weak acids such as formates are corrosive.

2. Adsorption Theory of Inhibition

One theory of inhibition proposed and developed by Brasher[13-15] proposes that the distinction between so called inhibitive anions (eg $[CrO_4]^{2-}$, benzoate) and aggressive anions (eg Cl^-, SO_4^{2-}) is, in effect, artificial. The theory suggests that all anions are aggressive at low concentrations but inhibitive at high concentrations. The difference in inhibitive 'ability' is that whereas some anions have a wide concentration range of aggressivity, others have a short concentration range, as follows:

	Aggressive Range	Inhibitive Range
Ion 1	$<1M$	$>1M$
Ion 2	$<0.001M$	$>0.001M$

On this basis, ion 1 would be considered as aggressive but ion 2 would be considered inhibitive. However, the designation is somewhat arbitrary. Inhibition is thought to occur by action on the anodic kinetics, favouring passivity. Aggressive ions, however, become incorporated into the film when it forms, reducing its protective properties. In practice, the concept is of limited use since there is no evidence either that successful inhibitors affect anodic kinetics in all cases, or that the uptake of aggressive species is uniform or even necessarily damaging.

3. General Precipitation Causing Inhibition

It has been suggested that inhibitors function by causing general precipitation of insoluble products. For example, iron does not corrode appreciably in concentrated H_2SO_4, due possibly to the formation of relatively insoluble $FeSO_4$.

Examples of specific insoluble film forming inhibitors are tannins and benzotriazole (BTA). Tannins inhibit iron corrosion by the formation of an insoluble surface iron tannate complex. Reaction occurs not with the iron surface but with the ferric (Fe^{3+}) ions produced by the corrosion process.[16] Benzotriazole specifically inhibits corrosion of copper and copper alloys by forming a stable and insoluble chelated surface compound with copper ions. It is preferentially adsorbed, forming a molecular film which strongly resists the diffusion of copper ions from the surface and so limits the corrosion rate.

In general, however, if the mechanism of inhibition was such that the formation of insoluble products alone prevented further corrosion, then a uniform uptake of inhibitor species would be anticipated. Furthermore, a high uptake of inhibitor would also be expected.

In practice, the evidence suggests that patchy uptake of inhibitor is often observed and that, overall, small quantities of it are taken up (often less than enough for a mono-layer), suggesting that the action of many inhibitors is more likely to be the repair of a natural film by a metal/inhibitor compound, rather than merely the formation of insoluble products. For example, Lumsden and Szklarska-Smialowska[17] observed three types of film formation behaviour after exposing iron to various inhibitor solutions, using ellipsometry and Auger electron spectroscopy to analyse the films.

Type (i) films (typified by the reaction of nitrites) were very protective but never contained inhibitor species. A large shift was found in the corrosion potential. This result was, perhaps, not surprising since the reduction of NO_2^- gives ammonia (at the film/solution boundary) which is not taken up by the oxidised iron film as it forms. However, the film formed is a passive film.

Type (ii) films (typified by the action of chromates) were less protective than type (i) and inhibitor was always detected in the film. In such cases, the resulting film comprised Fe_2O_3 and Cr_2O_3: the cathodic process in this example being:

$2CrO_4^{2-} + 10H^+ + 6e^- \rightarrow Cr_2O_3 + 5H_2O$

with the anodic oxidation of iron represented by:

$2Fe + 3H_2O \rightarrow Fe_2O_3 + 6H^+ + 6e^-$

A mixed product of two oxides is the net result.

Type (iii) films (typified by the precipitates produced in phosphate solutions) were the least protective and were very thick. In this case, a flaky precipitate of ferric phosphate was formed, since the solubility product for the compound had been exceeded.

With regard to the structure of protective films, Mayne and Page[18,19] have shown that radioactively labelled azelate uptake on iron is patchy, and other workers[20,21] have demonstrated by surface sensitive analytical techniques that the uptake of chromate species on aluminium surfaces is also patchy rather than uniform. The inferences are that some inhibitors go to certain sites that require inhibition, and that uniformity of uptake is not necessarily required for inhibition. These results reinforce the suggestion that repair of deficiencies in the passive film in many cases leads to inhibition.

INHIBITORS IN ACID SOLUTIONS

The corrosion of metals in acid solutions differs from that described above in that the corroding metal surfaces are oxide free. Consequently, substances which inhibit metallic corrosion in near neutral solutions do not generally inhibit corrosion in acid solutions.

Metallic corrosion in acid solutions can be inhibited by substances ranging from simple compounds such as halides or carbon monoxide to relatively complex organic compounds, many of which contain multiple bonds, aromatic ring structures and elements of Groups V or VI of the Periodic Table.

Two theories have been proposed to account for the inhibitive actions of all of these substances in acid solutions.[3] The first is the 'adsorption' theory which links the inhibitive properties of organic molecules with the physical shielding effects of adsorption films. The second is the 'energy' theory which considers that the adsorbed inhibitors affect the kinetics of the electrochemical reactions. However, neither of these theories adequately explains all the experimental evidence accumulated in the literature.

It is likely that two or more mechanisms operate simultaneously: actual inhibitor mechanisms probably lie somewhere between these extreme viewpoints.

ADSORPTION EFFECTS

The primary event of the inhibiting action of an organic compound is undoubtedly adsorption. Weak physisorption occurs due to the electrostatic attractive forces between ionic charges or dipoles on the inhibitor species and the electric charge on the metal surface. In addition, however, inhibitors can bond to metal surfaces by electron transfer to form a coordinate type of link, generally termed chemisorption. Although the distinction between these two types of adsorption is rather unclear, chemisorption is favoured by the presence of relatively loosely bound electrons, such as may occur in anions and in neutral molecules bearing lone pair electrons, or in electron systems associated with multiple bonds or aromatic rings.

The tendency towards stronger adsorption increases as the electrons become more loosely held (which occurs with decreasing electronegativity) in the order $O < N < S < P < Se$. It has also been possible in some studies[22,23] to relate the inhibitive properties of the molecules to their structural features. In the case of pyridine, for example, if a methyl substitution is made on the ring, the electron availability of the resulting picoline depends on the position of the substitution. The methyl substituent effectively pushes electrons away, which can in turn induce a conjugative movement of electron charge further round the ring structure. If the net result is increased electron availability, the inhibitive ability is enhanced. This can be illustrated as follows:

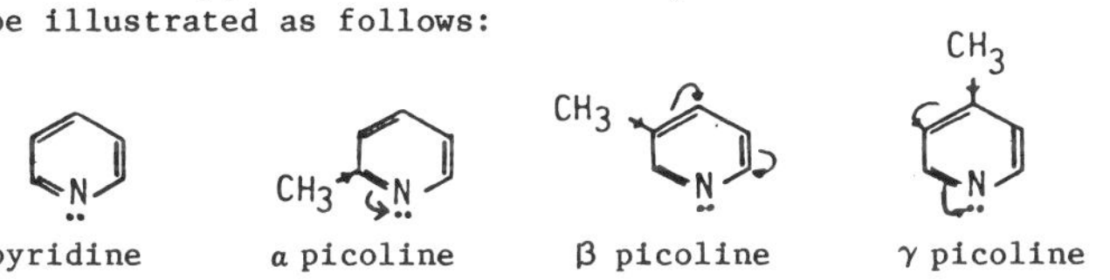

pyridine α picoline β picoline γ picoline

In this example, electron availability follows the trend:

pyridine < β picoline < α picoline ≈ γ picoline,

and the inhibition of iron in hydrochloric acid follows the trend:

γ picoline ≈ α picoline > β picoline > pyridine

These same principles can be applied to explain inhibitor efficiency trends in many other aromatic or indeed aliphatic organic molecules including, for example, amines and thiols. However, in addition to the availability of electrons at the reaction centre, there is a further requirement for electron accessibility. That is, substitution of progressively more bulky organic fragments will eventually prevent adsorption taking place because of steric hindrance. Thus, if the electrons of the reaction centre are surrounded by bulky fragments, they may be physically prevented from coming into close contact with the metal surface.

REACTION KINETICS EFFECTS

An organic inhibitor once adsorbed may retard anodic processes, cathodic processes, or both. In the latter case, the chemical would be termed a 'mixed' inhibitor. Again, changes in the corrosion potential when the inhibitor is added provide useful indications as to which process(es) are being retarded.[24] Surface films which physically restrict diffusion of metal ions are often formed, particularly with large molecules such as dextrin and large chain hydrocarbons, but the same effects are also observed with acetylenic compounds which form protonated alkynol complexes on ferrous based substrates.[25]

INHIBITION UNDER NON-AQUEOUS CONDITIONS

In atmospheric conditions which are not obviously wet, for example during storage, transportation or display, the need to mitigate the effects of metallic corrosion is often no less important. In practice, this is generally done in one of four ways:

1. de-humidifying agents
2. use of inert atmospheres
3. oils, greases, waxes and lacquers
4. volatile corrosion inhibitors

Although the first two and to some extent the third of these methods are really beyond the scope of the present paper, nonetheless it is worth noting that all four methods of limiting corrosion are in accordance with the principles rehearsed above. Drying agents, such as silica gel, can be used to lower the relative humidity of the air in an enclosed space. Mechanical air drying may also be used, as in air conditioning. In theory, it is possible to attain any level of relative humidity in an enclosed space, although the level required to ensure against corrosion depends on complex factors such as the composition of the atmosphere itself. Indeed, in the presence of some corrosive gases, a relative humidity level of less than 20% may be necessary. In all such circumstances, the limitation of corrosion is due to the lowered availability of cathodic reaction species (ie water and dissolved oxygen).

Oils, greases and lacquers function principally by interposing a physical barrier between the substrate and the corrosive species. In general, minimal chemical interaction occurs and the degree of protection depends mainly on the thickness of the coating applied. There are disadvantages to these measures of temporary protection. Oils and greases can undergo atmospheric oxidation and produce corrosive organic acids as side products, and lacquers may undergo degradation caused by ultra-violet light in the presence of oxygen and water, and so allow corrosion to proceed at damaged regions.

The fourth of the methods implies the use of volatile corrosion inhibitors or vapour phase inhibitors. Under these circumstances, vapourisation of selected compounds (the vapour pressures of which typically range from 10^{-2} to 10^{-7} mm Hg) enables inhibitors to reach the surface of the object to be protected. Inaccessible crevices and gaps can be thus protected. There is no need to dry the surrounding air, but the inhibitor and the objects to be protected must be properly sealed lest the volatile inhibitor escapes and no longer affords protection.

If the volatile compounds that are used in such circumstances are analysed, it is apparent that they all contain inhibitive species which have been discussed earlier. Typical volatile inhibitors are, for example, dibenzylamine nitrite and dicyclohexylamine chromate.[26] The mechanisms of inhibition are as discussed above. Amine species probably buffer the pH, whilst the nitrite or chromate probably act as oxidising agents promoting metal passivity.

The simultaneous protection of ferrous and non-ferrous metals by vapour phase inhibitors has proved to be a problem, since until recently the only compounds known to possess the property of passivating several metals were chromates, which in general are not volatile. However, in recent years, the development of multipurpose vapour phase inhibitors has yielded compounds such as dinitrobenzoates[3] which can adsorb onto and protect various ferrous and non-ferrous metals.

CONCLUSION

1. There are several ways by which metallic corrosion can be retarded. De-humidifying, de-oxygenating, adjusting pH levels or applying mechanical barriers all effectively reduce corrosion rates. The application of an inhibitor (a chemical added in small quantities to a corroding system) reduces metallic corrosion by interfering directly with the electrochemical mechanisms of corrosion processes.

2. In near-neutral solutions, most inhibitors are inorganic cations or anions which affect cathodic or anodic corrosion processes. These effects can be discussed in terms of a system's thermodynamics and kinetics. Simple

mechanistic explanations, however, cannot be given without reference to other factors such as pH buffering, competitive adsorption of aggressive species, precipitation of insoluble products and protective oxide formation/repair processes.

3. In acid solutions, it is predominantly organic molecules which inhibit corrosion by adsorption and protective film formation processes; but these effects cannot easily be separated from the system's reaction kinetics.

4. Mechanisms of metallic inhibition are complex and often are specific to one inhibitor/metal/environment system. However, some trends are discernable and these have made it possible to establish the principles of inhibitor mechanisms. As a consequence, it is possible to predict the types of chemicals (inorganic cations, anions or organic molecules) which should have some inhibitive efficiency. Predicted inhibition, however, may not be realised and in practice empirical screening of groups of candidate inhibitors is usually the method by which an inhibitor is selected for a particular purpose.

ACKNOWLEDGEMENTS

The author wishes to express his gratitude to his colleagues both in the Corrosion and Protection Centre Industrial Services and in the Centre itself, at UMIST, for assistance during the preparation of this paper. Particular thanks are due to Dr. G. Bailey and to Dr. J.D. Scantlebury for reading the manuscript.

REFERENCES

1. Putilova, I.N., Balezin, S.A. and Barannik, V.P., Metallic corrosion inhibitors, Pergamon, Oxford (1960).

2. Bregman, J.J., Corrosion inhibitors, Macmillan, London (1963).

3. Rozenfeld, I.L., Corrosion inhibitors, McGraw-Hill, USA (1981).

4. Pourbaix, M., Atlas of electrochemical equilibria in aqueous solutions, Pergamon, Oxford (1966).

5. Glasstone, S., An introduction to electrochemistry, Van Nostrand, Princeton (1942).

6. Bockris, J.O'M. and Reddy, A.K.N., Modern electrochemistry, Macdonald, London (1970).

7. West, J.M., Electrodeposition and corrosion processes, 2nd edition, Van Nostrand, London (1971), p. 29.

8. Tafel, J., Z. Fur. Physiki. Chemie. (1905), P. 641.

9. Evans, U.R., The Corrosion and oxidation of metals, Arnold, London (1960).

10. Evans, U.R., ibid, p. 128.

11. Hoar, T.P., Trans. Farad. Soc., 33, (1937), p. 1152.

12. Hoar, T.P., Association Belge pour l'Etude, l'Essai et l'Emploi des Materiaux, Brussels, June, (1954).

13. Brasher, D.M., Nature, 193 (1962), p. 868

14. Brasher, D.M., Br. Corros. J., 2 (1967), p. 95.

15. Brasher, D.M., Br. Corros. J., 4 (1969), p. 122.

16. Ross, T.K. and Francis, R.A., Corros. Sci., 18 (1978), p. 351.

17. Lumsden, J.B. and Szkarska-Smialowska, Z., Corrosion, 34 (1978), p. 169.

18. Mayne, J.E.O. and Page, C.L., Br. Corros. J., 7 (1972), p. 115.

19. Mayne, J.E.O. and Page, C.L., Br. Corros. J., 9 (1974), p. 223.

20. Abd. Rabbo, M.F., Richardson, J.A., Wood, G.C. and Jackson, C.K., Corros. Sci., 16 (1976), p. 677.

21. Koudelkova, M., Augustynski, J. and Berthou, H., J. Electrochem. Soc., 124 (1977), p. 1165.

22. Annand, R.R., Hurd, R.M. and Hackerman, N., J. Electrochem. Soc., 112 (1965), p. 138.

23. Annand, R.R., Hurd, R.M. and Hackerman, N., J. Electrochem. Soc., 112 (1965), p. 144.

24. Hoar, T.P. and Holliday, R.D., J. Applied Chem., 3 (1953), p. 502.

25. Tedeschi, R.J., Corrosion, 31 (1975), p. 130.

26. Knaack, D.F. and Brooks, D., Corrosion inhibitors, Ed. Nathan, C.C., NACE (1973), p. 221.

CORROSION INHIBITORS FOR CONSERVATION

S. Turgoose

A corrosion inhibitor may be defined[1] as "any substance which, when added in small amounts to the corrosive environment of a metal or alloy, effectively decreases the corrosion rate". Clearly, effective inhibitors could be a useful tool in the conservation of metals, and the possibility of their use has attracted a certain amount of interest. There have been several general discussions of the use of inhibitors in conservation[2-5] and others mention the possible inhibitive properties of compounds such as phosphates,[2-6] chromates,[7,8] nitrites,[7] tannins,[9] hydroxide[8,10,11] and carbonate.[12] A study has also been made of a range of organic inhibitors for acid cleaning[13] and volatile corrosion inhibitors are widely discussed but, it seems, rarely used. There is, however, one inhibitor in general use and that is benzotriazole. Its use in conservation has been reviewed[14] and while it was originally proposed[15] as a stabilisation method for copper alloys, largely because of its ability to form compounds with both copper(I) and copper(II), its inhibitive effect has been shown in some stripping solutions for copper alloys[16] and for iron.[13]

The above inhibitors were all developed for use on clean metal surfaces (ie those covered by at most a thin air-formed oxide film) and in some cases their inhibitive properties towards archaeological artefacts has been inferred from studies on such systems. Regrettably, little attention has been given to the particular problems that confront conservators and the way in which the nature of the artefact may influence the selection of particular inhibitors, or types of inhibitor, or may affect the decision as to whether to use inhibitors at all. Some of these problems are:

1. Many, probably most, excavated objects are covered by thick corrosion products and it is required to preserve these as well as any remaining metal core. The presence of these corrosion products can restrict inhibitor access to parts of the artefact and make inhibition more difficult.

2. It is required to preserve artefacts indefinitely and so in certain cases, particularly long-term storage, corrosion prevention is necessary, rather than simply a reduction in corrosion rate.

3. The appearance of the treated object is very important and this will generally rule out the possibility of incorporating inhibitors into multi-coat paint systems or other such opaque coatings.

The presence of composite objects (made from more than one material) and the practice of storage by site or by collection rather than by material, with its consequent environmental compromises, also make the use of inhibitors more difficult.

It is intended here to look at the possible application of corrosion inhibitors in conservation in the light of these particular problems and to assess which inhibitors or types of inhibitor may be useful in a given process with a given type of artefact. Before this can be done, it is necessary to know a little more about mechanisms by which corrosion inhibitors work.

THE ACTION OF CORROSION INHIBITORS

The corrosion of metals is electrochemical and the anodic and cathodic reactions may occur at different places on the object. The anodic reaction, metal dissolution (1) must obviously occur at the metal surface. This need not be true of the cathodic reaction, which is generally oxygen reduction (2) or hydrogen evolution (3) in acid solutions.

$$M \rightarrow M^{n+} + ne^- \quad \text{.... (1)}$$

$$O_2 + 2H_2O + 4e^- \rightarrow 4OH^- \quad \text{.... (2)}$$

$$2H^+ + 2e^- \rightarrow H_2 \quad \text{.... (3)}$$

The cathodic reaction can occur on the surface of any electronic conductor that is connected to the corroding metal, such as another metal, or graphite in the case of cast iron, or on a conducting corrosion product such as magnetite on iron.

There are several methods of classification of corrosion inhibitors but in terms of their application to heavily corroded material it is convenient first to draw the distinction between anodic and cathodic, according to whether they retard the anodic or cathodic reaction. For an inhibitor to be effective it must reach those electrode surfaces at which it can function, so anodic inhibitors must reach the metal surface and, indeed, all parts of the surface. Cathodic inhibitors, on the other hand, only need to reach those places to which the cathodic reactants can diffuse, and this may be easier. Inhibitors may be further classified as adsorption or film-forming, according to whether their action is due to an adsorbed layer or to a three-dimensional film. In general, the chosen inhibitors for near-neutral and alkaline solutions would be of the film-forming type, whereas adsorption inhibitors are generally used in acid solutions, although in both cases there are exceptions.

To consider first inhibitors for neutral and alkaline solution, comparison of the mechanisms of action of anodic and cathodic shows that there may be clear differences in their effectiveness, and thus usefulness, on corroded metal. The anodic ones function by formation of insoluble compounds with the metal ions produced at the anodic areas, and these precipitates may either assist in the maintenance of an oxide film on the metal surface, or may themselves form a film. The factors that influence the effectiveness of such inhibitors are:

1. the metal composition, since metal salts must be formed;
2. the inhibitor concentration;
3. the dissolved oxygen concentration;
4. the presence of other anions, such as chloride;
5. the solution pH.

In 2 - 5 the relevant conditions are those at the anodic areas, ie the metal surface, and the presence of corrosion products may prevent these factors from being under the control of the operator, since they will not be directly related to the bulk solution concentrations and pH.

Cathodic inhibitors generally function by virtue of their reaction with cathodically produced hydroxide ions and the consequent deposition of an insoluble product on the cathode surface. This film may restrict oxygen access to the cathode and lower the overall corrosion rate, since this is usually determined by the rate at which oxygen can be reduced. Unlike anodic inhibitors, access to the metal surface is not always necessary and the chemical nature of the cathode surface is often unimportant. Like anodic inhibitors, the efficiency of cathodic inhibitors depends on pH, since precipitation of a film must occur with only a small pH rise.

Another important difference between anodic and cathodic inhibitors is the effect when the concentration of inhibitor is too low to give complete protection. As mentioned above, the corrosion rate is generally under cathodic control and any film formation on the cathode surface will lead to a reduction in this rate. However, anodic inhibitors present in amounts too small to give complete protection do not reduce the corrosion rate at all, although some, particularly oxidising agents such as chromate and nitrite, may cause metal loss to be localised at small areas, with consequent deep pitting.

In acid solutions, adsorption inhibitors are generally used, partly because oxide films are soluble under such conditions and partly because one of their major applications is in the removal of oxide scale from metals and film formation would retard this process. These inhibitors are often nitrogen- or sulphur-containing organic molecules which bind to the metal surface via these atoms and retard either or both electrode reactions. Even the most efficient do not usually totally prevent corrosion, but the reductions in rate can be considerable and the problems of inhibitor access to the metal surface do not occur, since the treatment removes the corrosion products.

The factors that influence the effectiveness of inhibitors have been discussed in much more detail elsewhere,[17] but those discussed above are some that need to be borne in mind when considering their use in conservation, ie under circumstances rather different from those under which much of the original experimental work was carried out.

CORROSION INHIBITORS IN CONSERVATION

Before the decision is made to use an inhibitor as part of a conservation procedure, it is necessary to know whether corrosion is a problem and, if so, what can be achieved by preventing it or reducing its rate.

At the outset it should be stressed that many of the changes seen in metal artefacts after excavation are not directly due to corrosion of the remaining metal, and even if we were able to totally prevent further corrosion this would not necessarily result in a stable artefact. For example, the changes that may be seen in copper alloy objects, 'bronze disease', can be explained in terms of the oxidation of cuprous chloride by oxygen from the air in the presence of moisture, reaction (4), resulting in the formation of

$$4CuCl + O_2 + 4H_2O \rightarrow 2Cu_2(OH)_3Cl + 2H^+ + 2Cl^- \quad .... (4)$$

atacamite or paratacamite. Similarly, the deterioration of iron artefacts can be attributed to the oxidation of ferrous ions, reaction (5), and in neither of these cases will the

$$4Fe^{2+} + O_2 + 6H_2O \rightarrow 4FeOOH + 8H+ \quad .... (5)$$

reactions be prevented by stopping corrosion. However, in the absence of corrosion these reactions will cease when all of the air-unstable material present has been consumed, but if corrosion continues after excavation there is a continuing source of this material and the oxidation reactions will continue. Thus, while corrosion prevention will not guarantee stability, it will certainly assist, particularly in the many cases where only small amounts of unstable material are present. In the case of iron, the lamination of corrosion products may be caused by the very acidic conditions developed, due to reaction (5), and the consequent local dissolution of existing oxides. It would then follow that any chemical treatment that neutralised the acid produced would lead to greater stability, whether or not it reduced the corrosion rate. In the case of copper it is possible that reagents such as benzotriazole may react with CuCl to form a product that is less easily oxidised than cuprous chloride itself, and once again stability may be enhanced even without preventing corrosion. In general, treatment with a compound that is known to be a corrosion inhibitor for clean metal may have beneficial effects even if it does not act as such on corroded artefacts. While this does not detract from the usefulness of such treatments, there are times when it is important to know whether or not corrosion is continuing and to be able to predict whether a compound applied to a corroded metal artefact will prevent this. However, this is not an easy matter to decide.

The standard methods of measurement of corrosion rate and inhibitive properties are all based on studies on clean metal surfaces and the presence of corrosion products, or any film, limits their use. The appearance is not necessarily related to the presence or absence of corrosion and weight loss measurements are of little use, since other reactions may cause weight changes and the amount of included water may not be constant. Most of the electrochemical corrosion rate measurements rely on being able to measure the electrode potential of the metal, but this is not possible in the presence of thick corrosion products. One possible method of determining corrosion rates is to measure the consumption or production of a component of the cathodic reaction, and hydrogen evolution has been used for this purpose in an assessment of inhibitors for acid stripping solutions for use in conservation.[13] In theory, the rate of oxygen consumption could be used in a similar way in neutral and alkaline solutions, but the possibility of other reactions that also consume oxygen might make quantitative work difficult. There are perhaps a couple of electrochemical methods that might be applicable to corroded archaeological material. First, a modified polarisation resistance technique has been used for corroded steel (although probably very lightly corroded in conservation terms) in sea water,[18] and secondly, alternating current (a.c.) impedance methods, which have been used on metals covered by paint films, might have some application, although they have not yet been applied to archaeological material.

We are thus in the position that even if we wish to use a corrosion inhibitor we cannot, in many cases, measure the action that it has in use, and inhibitors have to be selected on the basis of their known properties and our assessment of the type of inhibitors that might be successful, rather than on the results of short-term tests on corroded metal. Further, in the cases where 'corrosion inhibitors' have been used for archaeological material, there is rarely, if ever, any proof that they do in fact behave as such on corroded metal. Clearly, when dealing with clean metal surfaces and with inhibitors for stripping solutions, these problems do not occur, but these are perhaps minor examples of the potential uses of inhibitors in conservation.

The selection of an inhibitor for use in a conservation procedure is clearly dependent on many factors, such as the composition of the solution or environment, the allowable corrosion rate, the metal composition and the possible presence of other materials. Along with these, the possible use of other methods of corrosion control must be borne in mind, since it will not always be true that an inhibitor is the best method, or indeed even a possibility. With all this in mind, it is now intended to discuss the application of inhibitors in particular stages of conservation and it is convenient to consider two categories: solution treatments, which last for a limited time, and storage, where solutions are not generally involved, at least not intentionally, and the time scale involved is indefinite.

SOLUTION TREATMENTS

Two very different processes are included here, washing and stripping. In the former, near-neutral or alkaline solutions are used and the aim is to remove soluble salts from corroded artefacts without removing the corrosion products, whereas in the latter the intention is to remove the corrosion products.

Washing is more commonly used for iron objects than for other metals, and ions such as hydroxide, chromate and nitrite have been claimed to have inhibitive effects when used in wash solutions. While these are all perfectly good inhibitors for clean iron, they are all anodic inhibitors, and this must cause doubts as to whether they will exert any inhibitive action on corroded metal. There are several reasons for this:

1. access of inhibitor to all parts of the metal surface is probably impossible. After all, long washing times are often necessary because of the very slow equilibration of chloride between metal surface and bulk solution, and there is no reason to suppose that inhibitive anions

will move through the corrosion product layers with any greater ease. Apart from this,

2. the high chloride levels that will initially exist at the metal surface will make inhibition more difficult, and

3. the pH at the metal surface is initially quite low and this must often be raised before the inhibitors will be effective.

The overall effect of these difficulties is to make it extremely unlikely that anodic inhibitors will have any beneficial effects in washing unless they arise simply from chemical reactions of the substance. In this context it may well be that the lack of visible deterioration in artefacts treated in sodium hydroxide solution is due to the prevention of acidity, which would otherwise build up due to reaction (5). Further, the results of continuing corrosion will not be seen if the initial anodic products, ferrous ions in this case, are precipitated within the existing corrosion product layers, as will happen if these layers are made alkaline.

An alternative would be to use cathodic inhibitors which do not necessarily have to reach the metal surface, but only those places to which oxygen can diffuse. If the cathodic areas are an appreciable distance from the metal surface, as often appears to be the case with iron, this may be much easier. Also, even if there is not inhibitor access to all parts of the cathodic areas, some corrosion rate reduction may be obtained. This type of inhibitor is less well documented as regards use in neutral and alkaline solution, probably because they are generally less effective than anodic for clean metal. Film-forming cathodic inhibitors include metal ions such as Zn^{2+} and Mg^{2+}, which precipitate insoluble hydroxides at the cathodic areas, and polyphosphates such as "sodium hexametaphosphate" with calcium ion additions. The properties of the latter suggest they might have some application as inhibitors on archaeological iron, not least the fact that they are effective on corroded surfaces,[19] and should cause no damage to other metals or organic materials. They deposit a film of uncertain composition[20] containing calcium, phosphate and perhaps iron, although films can be formed on inert cathodes, such as graphite and platinum.[21] In a study[22] of sodium hexametaphosphate-calcium solutions for the washing of iron, chloride extraction rates have been found to be comparable to or better than sodium hydroxide solutions, and no visible change to the artefacts occurred. The problems, already mentioned, of assessing inhibitive effects on corroded surfaces do not, however, permit it to be claimed with certainty that the solution is inhibiting corrosion.

Cathodic adsorption inhibitors for neutral and alkaline solutions have hardly been studied, although they are widely used in acidic solutions. In this case, it is the cathodic evolution of hydrogen on a bare metal surface that is retarded, whereas in neutral solutions the reduction of oxygen, perhaps on an oxide surface, is the cathodic reaction, so different compounds may be required. However, once again, these compounds would not be expected to give complete protection and the possible benefits that may accrue from their use must be compared to those that can be obtained by other methods of corrosion control.

In washing, the use of coatings is clearly impossible, as is the removal of water (if the use of non-aqueous solvents is discounted, since they seem much less effective than aqueous systems[23,24]), but it is possible to remove oxygen from the solution and this will reduce the corrosion rate to an imperceptible level. It is not experimentally difficult as long as sealed containers are available and one washing solution already in use (alkaline sulphite[25]) does remove oxygen from the solution. It was not the reason that the method was originally proposed and oxygen removal can easily be accomplished in less extreme conditions by, for example, use of dilute, approximately neutral, sulphite solutions with, if desired, the addition of a few parts per million of a metal ion such as Co^{2+} to catalyse the reaction with oxygen. If corrosion is perceived as a major problem in washing then this course, oxygen removal, would seem to be a better way of preventing it than the use of inhibitors, but the latter are perhaps easier to use and may give some reduction in corrosion rate, although, unlike oxygen removal, they will not prevent the oxidation of existing corrosion products.

The use of acidic solutions for the removal of corrosion products is one case where much of the fundamental work has a direct relevance to archaeological metals, since the desired reactions are the same as in the pickling of metals to remove oxide scales. Inhibitors for these solutions are almost exclusively of the adsorption type and rarely give complete protection, which must be remembered when their use is considered. For inorganic acid solutions (eg sulphuric and hydrochloric acids) they are usually nitrogen- or sulphur-containing organic molecules, such as quinoline and thiourea and their derivatives, and their effectiveness depends on the metal composition and the nature of the acid used. These factors mean that it is difficult to predict the effectiveness of a corrosion inhibitor in one stripping solution from results obtained in another, especially if the metal composition is also different. The problems can be illustrated by reference to the use of benzotriazole in the stripping of copper alloys. In sulphuric acid it is more effective for copper than for brass,[26] and in other stripping solutions Merck[16] showed that the dissolution of tin from tin-bronzes is inhibited in alkaline solutions such as tartrate and glycerol but not in citric acid. Thus inhibitors must be tested under the conditions of use before their effectiveness can be guaranteed.

Wilthew has studied some inhibitors, including benzotriazole and thiourea, for use in sulphuric or hydrochloric acid stripping solutions for iron,[13] and compared these solutions to the more commonly used citric and phosphoric acids. It was found that although metal corrosion rates were slightly higher in inhibited sulphuric acid than in citric, this was compensated for by the shorter treatment times involved, and the amount of metal loss during stripping was less in the sulphuric acid. This suggests that there is a considerable potential use for inhibitors when stripping is required. There is much less information about possible inhibitors for near-neutral and alkaline stripping solutions but, again, they could be useful if effective.

STORAGE

The problems of corrosion control during storage are enhanced by the need for an indefinite lifetime for artefacts, so no corrosion rate can be deemed acceptable. The usefulness of anodic and cathodic inhibitors is determined by the factors already discussed and, in view of the difficulty of obtaining corrosion prevention, particularly on corroded surfaces, by the use of inhibitors alone, it would seem that other methods of corrosion control must be used. These may, however, be to supplement the use of inhibitors rather than instead of them. Practicality would dictate that immersion in solution is unacceptable in the long term, although it may be feasible for some objects between excavation and conservation, so environmental control is limited to conditioning the atmosphere. Inhibitors may be applied to artefacts in several ways; in coatings, by impregnation or dipping, and by transport in the vapour phase.

The full benefits of using inhibited coatings will only be realised if the coatings are applied to a perfectly clean surface, when they assist in the reduction of corrosion rate by virtue of their high electrolytic resistance. This is not possible on corroded metal and the influence of the coating is limited to providing a means of application of the inhibitor and perhaps of increasing the mechanical strength of the corrosion product layers. Whilst the permeability, to oxygen and water, of all the coatings likely to be used is high enough not to reduce corrosion rates, coatings used in conjunction with dry storage may also prevent condensation of water on the metal surface if there are short-term defects in the control system. The disadvantages of coatings largely lie in the effects they may have on the appearance of the object and in the time taken to apply them, particularly since no coating is likely to have an indefinite life and removal and recoating is almost certain to be necessary at some stage. Except for benzotriazole in an acrylic laquer, this method of inhibitor application has not been widely used, although a zinc phosphate containing primer has been applied to iron[6] and 4,4'-diaminodicyclohexylmethane has been used in coatings on lead.[27] The former, being a mixed inhibitor affecting both anodic and cathodic reactions, may have some effect on corrosion rate on corroded surfaces.

The impregnation of a corrosion inhibitor into a corrosion product layer has been widely used in recent years in the case of benzotriazole (BTA) for copper base artefacts. This certainly appears to be a successful way of stabilising such objects, but it is far from clear to what extent this is due to BTA acting as a corrosion inhibitor as opposed to being simply a chemical reagent. BTA is known to form compounds with both copper(I) and copper(II) and, to some extent, the increased stability could be due to the modification of the oxidation reaction (4) rather than to corrosion prevention. If we wish to develop other inhibitors for archaeological material it is important to know to what extent the various properties, of those currently used, contribute to stabilisation, so that we can assess whether compounds such as 2-aminopyrimidine, which is reported[28] to be as effective as BTA as a corrosion inhibitor, are likely to be useful in the treatment of copper alloy objects.

Dipping can be used to apply inhibitors to clean metal surfaces, and a variety of anti-tarnish agents for both silver and copper can be used in this way. Many of the patented compounds are of the general formula R-SH, although nitrogen-containing compounds (imidazole derivatives) can be used for copper. Most of these will not indefinitely prevent tarnishing but may reduce the rate at which it occurs. An important difference between corroded surfaces and clean metal surfaces is that in most cases the former will already contain corrosion stimulants, whereas the latter corrode due to the uptake of such stimulants from the atmosphere, and environmental control can remove many of the corrosion problems of clean metal surfaces.

The use of volatile corrosion inhibitors in conservation has been discussed,[4] but they rarely seem to be used. They possess sufficient vapour pressure to move in the vapour phase and so can reach all parts of an artefact's surface without the need for an immersion treatment, but there is no reason to suppose that they act in a different manner from other inhibitors, ie their effect is to dissolve in condensed water on the surface and give an inhibitive solution.[29] A large range of compounds can work in this way[30] but the most common are amine nitrites, carboxylates, chromates or carbonates. The latter are, in fact, usually carbamates and the commonly used "cyclohexylamine carbonate" is an adduct of cyclohexylamine and carbon dioxide, cyclohexyl ammonium cyclohexylcarbamate $[C_6H_{11}.NH_3^+][^-OOC.NH.C_6H_{11}]$, although the transport in the vapour phase is as separate cyclohexylamine and carbon dioxide.[31] This and many amine nitrites are effective on clean iron surfaces, but one major problem with their use is that they increase the corrosion rate of many non-ferrous metals.[32] In addition to this, many are, in varying degree, toxic[33] and must be used in sealed containers to prevent depletion of inhibitor. In considering their use against other methods of corrosion control, this latter factor is important, since once a closed container is available other methods of environmental control become possible. In particular, the tarnishing of silver can be prevented by removal of sulphur-containing gases from the atmosphere, and lead corrosion is generally caused by the presence of organic acids. Also, in dealing with corroded metals, impregnation of the corrosion products may be a better way to ensure inhibitor availability over the whole artefact surface, since the types of inhibitor that could be used would not be restricted by vapour pressure considerations.

CONCLUSIONS

The usefulness of corrosion inhibitors in conservation is limited by the nature of the objects that must be treated, particularly the occurrence of corrosion products that must be preserved. There are also problems in determining inhibitive effects on metals covered by corrosion products, especially since it is difficult to separate the inhibitive effects from the chemical reactions of the "corrosion inhibitor", but on the basis of our knowledge of the mechanisms by which various inhibitors work we can predict which types of inhibitor may be useful in various circumstances.

The difficulty of inhibitor access to the metal surface means that anodic inhibitors are unlikely to be effective and cathodic inhibitors, which may work, do not generally give complete protection. In stripping and washing solutions some corrosion may be allowable, but in long-term storage corrosion prevention must be the aim and this will rely upon the use of some form of environmental control, such as the removal of water or atmospheric pollutants. In this, the use of a corrosion inhibitor might provide an additional safeguard in the event of a failure of the control.

In general, corrosion inhibitors should be considered as one possible method of corrosion control amongst many others, and they should all be considered in the assessment of the optimum treatment for a given artefact.

ACKNOWLEDGEMENTS

The author would like to thank the SERC for financial support and Prof. R.D. Gillard and Dr. D. Leigh for their encouragement.

REFERENCES

1. Eldridge, G.G. and Warner, J.C., H.H. Ulig (ed), The corrosion handbook, (1948), p. 985.

2. Fenn, J.D. and Foley, K., Passivationof iron, Conservation in Archaeology and Applied Arts, IIC Congress, Stockholm (1975), pp. 195-8. IIC, London.

3. Foley, R.T., Measures for preventing corrosion of metals, Corrosion and metal artefacts, NBS Special Publ. 479 (1977), pp. 67-75.

4. Stambolov, T., Corrosion inhibitors, ICOM Committee for Conservation, 5th Triennial Meeting, Zagreb (1978), 78/23/9 1-6.

5. Walker, R., The role of corrosion inhibitors in the conservation of iron, Conservation of Iron, National Maritime Museum monograph no. 53 (1982), pp. 58-65.

6. North, N.A. and Pearson, C., Investigations into methods of conserving iron relics recovered from the sea, Conservation in Archaeology and the Applied Arts, IIC Congress, Stockholm (1975), pp. 173-81, IIC, London.

7. North, N.A. and Pearson, C., Methods for treating marine iron, ICOM Committee for Conservation, 5th Triennial Meeting, Zagreb (1978), 78/23/3, pp. 1-10.

8. Hamilton, D.L., Conservation of metal objects from underwater sites: a study in methods, Texas memorial Museum (misc. paper no. 4) and Texas Antiquities Committee (publ. no. 1) (1976), pp. 21-25.

9. Pelikan, J.B., Conservation of iron with tannin, Studies in Conservation, 11 (1966), pp. 109-14.

10. Pearson, C., The preservation of cannon after 200 years under the sea, Studies in Conservation, 17 (1972), pp. 91-110.

11. Fabech, E.W. and Trier, J., Notes on the conservation of iron, especially the heating to redness and the lithium hydroxide method, ICOM Committee for Conservation, 5th Triennial Meeting, Zagreb (1978), 78/23/13, pp. 1-6.

12. Oddy, W.A. and Hughes, M.J., The stabilisation of active bronze and iron by the use of sodium sesquicarbonate, Studies in Conservation, 15 (1970), pp. 183-9.

13. Wilthew, P.T., Inhibited sulphuric and hydrochloric acid stripping solutions, ICOM Committee for Conservation, 6th Triennial Meeting, Ottawa (1981), 81/23/6, pp. 1-8.

14. Sease, C., Benzotriazole: a review for conservators, Studies in Conservation, 23 (1978), pp. 78-85.

15. Madsen, H.B., A preliminary note on the use of benzotriazole for stabilising bronze objects, Studies in Conservation, 12 (1967), pp. 163-6.

16. Merck, L.E., The effectiveness of benzotriazole in the inhibition of the corrosive behaviour of stripping reagents on bronzes, Studies in Conservation, 26 (1981), pp. 73-6.

17. Mercer, A.D., Corrosion inhibition: principles and practice, Corrosion, (ed L.L. Shreir) 3rd edition (1976), pp. 18:9-18:33.

18. Azerri, N., Bruno, R. and Splendorini, L., Assessment of corrosion rate of steel in sea water by polarisation resistance technique, Corros. Sci., 21 (1981), pp. 781-7.

19. Evans, U.R., The Corrosion and Oxidation of Metals, 1st suppl. vol., Edward Arnold (1976), p. 76-77.

20. Evans, U.R., The Corrosion and Oxidation of Metals, 2nd suppl. vol., Edward Arnold (1976), p. 99.

21. Hatch, G.B., Protective film formation with phosphate glasses, Ind. Eng. Chem., 44 (1952), pp. 1775-80.

22. Turgoose, S., unpublished results.

23. North, N.A. and Pearson, C., Washing methods for chloride removal from marine iron artefacts, Studies in Conservation, 23 (1978), pp. 174-86.

24. Watkinson, D., An assessment of lithium hydroxide and sodium hydroxide treatments for archaeological ironwork, Conservation of Iron, National Maritime Museum monograph no. 53 (1982), pp. 28-40.

25. North, N.A. and Pearson, C., Alkaline sulphite reduction treatment of marine iron, ICOM Committee for Conservation, 4th Triennial Meeting, Venice (1975), 75/13/3, pp. 1-13.

26. Walker, R., Triazole, benzotriazole and naphthotriazole as corrosion inhibitors for brass, Corrosion, 32 (1976), pp. 414-7.

27. Griz, V.E., Milova, L.Ya., Kuchinskii, V.N., Levin, S.Z. and Vinogradova, V.P., Inhibited coatings for the protection of lead archaeological objects, for storage in the State Hermitahe, Prot. Met., 12 (1976), pp. 180-2.

28. Horner, L. and Pliefkr, E., Inhibitors of corrosion 28: 2-aminopyrimidine as an inhibitor of the corrosion of copper in salt solutions under oxygen, Werkst. Korros., 33 (1982), pp. 189-93.

29. Pourbaix, M., Discussion in Corrosion and Metal Artefacts, NBS Special Publ. 479 (1977), pp. 223-4.

30. Baker, H., Volatile rust inhibitors, Ind. Eng. Chem., 46 (1954), pp. 2592-8.

31. Henriksen, J.F., The distribution of a volatile corrosion inhibitor (CHC) on corroded iron, Corros. Sci., 12 (1972), pp. 433-6.

32. Miksic, B.A., as ref. 29.

33. Paustovskaya, V.V., The toxicity of atmospheric corrosion inhibitors, Prot. Met., 10 (1974), pp. 310-1.

SKERRY/TURGOOSE DISCUSSION

Andrew Agyrakis Will corrosion inhibitors affect subsequent analysis of the object?

Stephen Turgoose If the inhibitors were metal ions or metal-containing anions then they may show up in subsequent analysis. The addition of calcium probably would not cause confusion but the undocumented addition of, say, chromium, from chromate might. Organ (Corrosion and metal artefacts, NBS Special Publication no. 479, 1977, p. 113) tells a story of Roman bronze "containing" chromium, which in that case arose from stainless steel during electrolysis, and similar occurrences could arise.

Janey Cronyn I'm slightly confused about the chloride ions. What is the result of having chloride ions around?

Brian Skerry Chloride ions, like most anions and cations, have complicated reactions. The presence of chloride, generally, has a destructive effect on the passive film which would otherwise inhibit corrosion. Following adsorption, chloride ions can penetrate and disrupt an otherwise coherent oxide film, causing the promotion of corrosion processes at the metal surface. However, it is an empirical result that, at least for aluminium alloys, if present in sufficient concentrations, chloride ions may suppress corrosion rates. One explanation for this observation could be that high concentrations of adsorbing chloride ions may physically restrict the flow of oxygen to cathodic sites.

Stephen Turgoose In near-neutral solutions the presence of chlorides and many other aggressive anions, such as sulphate and nitrate, can reduce the effectiveness of inhibitors. The higher the concentration of chloride, the higher the inhibitor concentration required to prevent corrosion. It appears to be related to competitive adsorption of the various ions on the metal surface, and if the inhibitor cannot be adsorbed it is not effective.

Todor Stambolov What kind of volatile corrosion inhibitors have you used?

Stephen Turgoose I have looked at the cyclohexylamine based ones; the carbonate, so-called, and the benzoate. The benzoate is a bit better for non-ferrous metals, especially copper where the carbonate causes problems. But I have worked with aqueous solutions because I don't think there is any reason to suppose that they function in any manner other than by dissolving in the water on the surface. Cyclohexylamine carbonate is just as good an inhibitor as sodium carbonate solution of the same pH as you get. So the cyclohexylamine is conditioning the pH and the carbonate is the inhibitor.

Brian Skerry The real advantage of volatile corrosion inhibitors is that the inhibitive species diffuse through the atmosphere to the surface to be protected and reach into cracks and other fairly inaccessible areas. However, you need to take some care in choosing a volatile corrosion inhibitor, especially if a mixed metal system is to be protected. For example, amines may inhibit steel while accelerating copper corrosion. Environmental control is also very important; volatile inhibitors must only be used in enclosed spaces.

Kate Hunter You mentioned the breakdown of lacquers with ultra violet. Could you suggest which might be the best to use?

Brian Skerry All lacquers will break down under natural uv radiation unless they have an uv absorbent incorporated into the polymer. There are many types of organic lacquer and a considerable number of uv absorbers. Lacquers do not act directly as corrosion inhibitors, they merely interpose a physical barrier between the environment and the metal substrate. Lacquers tend to become brittle when they degrade. If the metal substrate becomes exposed where the lacquer has broken down, localised and accelerated corrosion attack may occur.

Stephen Turgoose Lacquers and coatings generally are difficult to apply perfectly and if, for example, you are trying to coat silver it may well be that aesthetically a slow general tarnish is more acceptable than a mottling. It takes time to apply lacquers as well, so you have to weigh the benefits.

BENZOTRIAZOLE: A PERSPECTIVE

H. Brinch Madsen

For over 30 years, benzotriazole (BTA) has been successfully employed as a corrosion inhibitor for copper and copper alloys,[1] in particular to prevent corrosion due to environmental conditions during packing, storage and transport. The main application, however, is to protect copper and its alloys by immersing them in solution.[2,3] In 1967, benzotriazole was adopted for the treatment of antiquities of copper and copper alloys;[4] since then, it has been commonly employed in conservation laboratories. The role of BTA in the stabilisation of copper and alloy antiquities, and the relevant literature, have been reviewed by Catherine Sease,[5] R.D. Bleck[6] and R. Walker[2,3] among others.

Benzotriazole, $C_6H_5N_3$, is a white to light tan, odourless powder with a melting point of 95°. It is a heterocyclic compound with the following structural formula (Fig. 1):

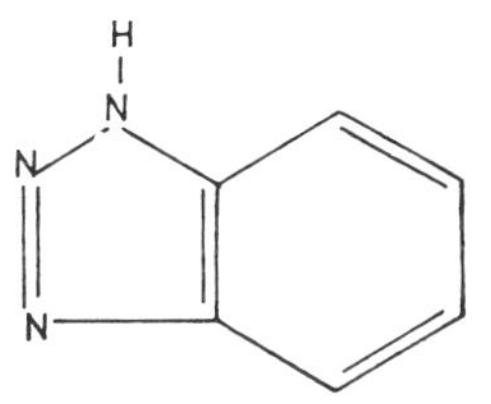

Fig. 1 Structure of benzotriazole

When BTA is in contact with copper and copper alloys, it is considered to form a thin, chemically bonded complex, cuprous benzotriazole, on the surface, where it reinforces the protection already afforded by the naturally occurring oxide film.[2] It has been shown that this process is rapid on a cuprous oxide layer but very slow on a cupric surface.[3]

The thickness of the copper-benzotriazole film has been given as 40-140 A°, although values of up to 5000 A° have been reported.[3] The surface film produced may be a multilayer consisting of copper-cuprous oxide-chemisorbed cuprous benzotriazole/anodic corrosion product/cuprous benzotriazole.

The copper-benzotriazole molecule is thought to be produced by the substitution of a cuprous ion for a hydrogen ion in the nitrogen-hydrogen group.

Two possibilities are proposed for the linking of these molecules:

1. A cuprous benzotriazole complex on the surface consisting of polymeric chains (Fig. 2);

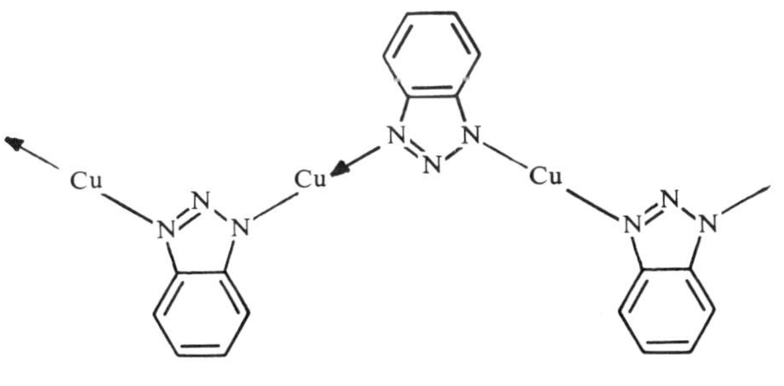

Fig. 2 Cuprous benzotriazole complex

2. A cupric benzotriazole complex which is possibly a network structure (Fig. 3).

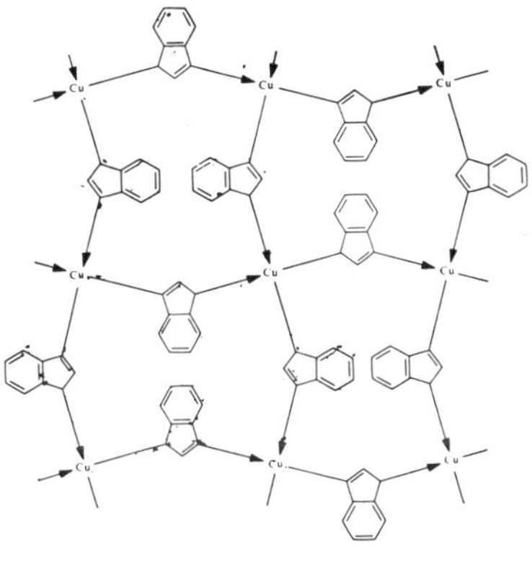

Fig. 3 Cupric benzotriazole complex

Several studies suggest that the cuprous complex is the more probable.[3]

There have been a considerable number of studies on the inhibition of copper corrosion by benzotriazole in neutral and acidic chloride solutions.[7] BTA adsorbs onto copper and its oxide surfaces from chloride solutions to form a surface film. Numerous investigations have been carried out into the chemical nature, composition, thickness and structure of this surface complex, using such varied techniques as infra-red reflectance, spectroscopy, ellipsometry, X-ray photo-electron spectroscopy and electrode impedance.[8]

The inhibition in chloride solution has been explained by Evans[9] as formation of a basic copper chloride which interacts with BTA to form a blocking substance. The inhibitive action of benzotriazole is attributed to the formation of an impervious Cu-BTA film on the surface.

The impedance studies of the Cu-BTA film formed in sodium chloride solutions indicated that the corrosion of copper is necessary for cuprous ions to participate in the formation of the protective film. The thickness of the film increases with the BTA concentration, and the concentration of BTA has to be high enough to reduce the dissolution rate of the Cu-BTA film and to repair any cracks in the film. The corrosion rate decreases with increasing film thickness. This fact may indicate that the slow dissolution rate of the film and slow transport of cuprous ions through it are responsible for the efficiency of the film as an inhibitor.[10]

This suggests that the film formed on copper in the presence of BTA has features similar to those of a passivating film on transition metals or alloys.[11] The organo-metallic complex precipitated on the surface of the metal thus produces a barrier to the corrosive environment.[12]

On the basis of the earlier results of Cotton and co-workers with BTA as inhibitor for copper[13,14,15] I attempted in 1967 to apply BTA treatment to corroded prehistoric bronzes. Their susceptibility to corrosion was demonstrated by placing the objects in a humidifier, where the relative humidity was near saturation point, for 24 hours.

The objects were carefully cleaned with equal parts of acetone and toluene and then immersed in a 3% by weight solution of BTA in alcohol. The immersed object was placed in a vacuum until no more bubbles emerged from it. After this treatment, the whole procedure was repeated, until after one week no further visible signs of corrosion were found. The results of this treatment were published in Studies in Conservation 1967. The method has subsequently been widely applied to the stabilisation of both prehistoric and recent bronzes. 17 years have now passed and thousands of objects have been treated with BTA.

BTA treatment is a relatively inexpensive and simple method, the effect of which may be readily assessed in a humidity chamber. Confidence in the method is also supported by the considerable research which is being carried out on benzotriazole as a copper corrosion inhibitor. This has already documented that BTA is effective and several reports give us an increasingly better idea of what occurs, but in these cases we are dealing with well-known materials and environments which may be monitored with sophisticated equipment. Museum objects are in a different category, and in this case we seldom know anything about the composition of the alloy. The objects have corroded under a variety of conditions, are coated with a complex of corrosion products and are exhibited or stored in rooms whose ambient chemical and physical conditions we know nothing of, perhaps with the exception of the relative humidity. We, nevertheless, treat all copper alloy objects alike and expect or at least hope that they have been stabilised for ever. But are we right? Or what can we do to assure ourselves that our treatment works, also in relation to the time factor? In other words, is there any perspective in treating bronze with benzotriazole?

In order to answer this question or approach an answer, I have elected to examine the state of preservation of objects treated earlier.

Scrutiny of the records of the Conservation Department of the National Museum pertaining to the period 1967 to 1976 has revealed that over this 10 year period more than 500 objects have been treated with benzotriazole. The procedure has been practically the same in each case: after cleaning, the objects have been placed in a 3% solution of BTA in alcohol and transferred to a vacuum tank until no further bubbles appeared. They have been varnished with either Incralac or Bedacryl.

From the treated bronzes I selected all those which were in the collections of the National Museum and which had received no conservation prior to treatment with BTA. Of the 320 objects available for examination, 317 showed no signs of corrosion, while three had one or two small light green spots.

During the same period, fifteen Egyptian bronzes and a bronze sword from Amoryos, Greece were treated by Antiksamlingen.

Similarly, the conservation records of Forhistorisk Museum, Moesgard, Arhus, pertaining to the period 1969 to 1973 were examined. Here, 278 prehistoric and medieval Danish copper alloy objects treated with BTA according to the same procedure, with the difference that they had all been varnished with Incralac, were available for study. They were all stable. Orientalsk Samling there yielded 27 bronzes treated with BTA and available for study. They were all stable.

I thus had the opportunity to scrutinise 641 bronzes, only three of which, or 0.45%, had indications of corrosion.

This result should naturally be treated with all possible caution. Among other things, there is no record of whether all the bronzes were corroded before treatment or whether the three corroded specimens had been examined for corrosion after treatment. Nor has the condition of treated objects been compared with that of untreated ones. I also lack information on the relative humidity in the various storage areas. Nevertheless, I believe that the investigation supports the view that BTA is a good inhibitor of corrosion in copper alloys.

However, I also have to report that I have seen bronzes treated with BTA which have revealed corrosion after a number of years, namely bronzes exhibiting strong chloride corrosion, like a sheet bronze from Bahrain, which has corroded in a pergamyn bag.

In this case, I believe the corrosion is due to the bag having worn down the thin organo-metallic film, affording entry to air and damp.

Finally, I looked at the first BTA-treated bronze, namely the mirror from Egypt, the treatment of which I published in 1967. It was actively corroding before treatment, but afterwards stayed in a humidity chamber for three years without anything happening. After this it was varnished with Incralac and has now been in the collection for 14 years.

In connection with this paper, I was permitted to remove it from the collection and test its stability. I stripped the varnish and placed the mirror in a humidity chamber. After 24 hours I returned and, to my great surprise, found a multitude of small dark green spots on the surface, which quickly dried into dark green solid elevations as I examined them under the microscope.

These spots are now being examined at the University of Copenhagen. I can only surmise what they are and what has occurred and eagerly await the results of the examination.

I was given the title Benzotriazole: a perspective. I have tried to follow it in undertaking an investigation of previously treated objects. The statistical analysis was very encouraging, but the corrosion phenomenon on the little mirror calls for intensive research and this aspect will have to be thoroughly investigated before I care to pronounce on the perspective in benzotriazole treatment of corroded copper alloys.

REFERENCES

1. Proctor and Gamble Ltd., British Patent 652, 339, December 1947, London.

2. Walker, R., Benzotriazole as a corrosion inhibitor for immersed copper, Corrosion, NACE, 29 (7) (1973).

3. Walker, R., The role of benzotriazole in the preservation of copper-based antiques, The conservation and restoration of metals. Proceedings of the symposium, Edinburgh 1979, Scottish Society for Conservation and Restoration, Edinburgh.

4. Brinch Madsen, H., A preliminary note on the use of benzotriazole for stabilising bronze objects, Studies in Conservation, 12 (1967).

5. Sease, C., Benzotriazole: a review for conservators, Studies in Conservation, 23 (1978).

6. Bleck, R.D., Korrosionsinhibitoren in der Konservierungspraxis I. Benzotriazol, New Museumskunde, Jahrgang 18 (2) (1975).

7. Chin, R.J., Altura, D. and Nobe, K., Corrosion inhibition of iron and copper in chloride solutions by benzotriazole, Corrosion, NACE, 29 (5) (1973).

8. Lewis, G., The adsorption of benzotriazole onto cuprous oxide surface. An electrical impedance study, Corrosion NACE, 34 (12) (1978).

9. Evans, U.R., The corrosion and oxidation of metals, first supplementary volume, London (1968).

10. El-Taib Heakel, F. and Haruyama, S., Impedance studies of the inhibitive effect of benzotriazole on the corrosion of copper in sodium chloride medium, Corrosion Science, 20 (1980).

11. Mansfield, F., Smith, T. and Parry, E.P., Benzotriazole as corrosion inhibitor for copper, Corrosion NACE, 27 (7) (1971).

12. Fox, P.G., Levis, G. and Boden, P.J., Some chemical aspects of the corrosion inhibition of copper by benzotriazole, Corrosion Science, 19 (1979).

13. Cotton, J.B., Control of surface reactions on copper by means of organic reagents, Proceedings of the second international congress on metallic corrosion, National Association of Corrosion Engineers, Houston, Texas, New York (1963).

14. Cotton, J.B. and Scholes, I.R., Benzotriazole and related components as corrosion inhibitors for copper, British Corrosion Journal, 2 (1967).

15. Dugdale, I. and Cotton, J.B., An electrochemical investigation on the prevention of staining of copper by benzotriazole, Corrosion Science, 3 (1963).

TESTING ANTI-TARNISH PREPARATIONS

S.M. Bradley

The tarnishing of silver by hydrogen sulphide, H_2S, evolved from textiles and other materials used in the storage and display of antiquities has been investigated by Oddy, Daniels and Blackshaw at the British Museum.[1,2,3] This work is still continuing but the results obtained so far show that the rate at which silver objects tarnish in museums can be reduced by careful selection of materials for the construction of display and storage areas. However, because a background level of hydrogen sulphide amounting to 0.3Mg m^{-3} (0.19 ppbv) is present in the atmosphere as a pollutant, tarnishing cannot be completely prevented. There is also evidence to suggest that carbonyl sulphide (OCS) may also tarnish silver. Carbonyl sulphide concentrations of 0.07 to 0.37 ppbv have been measured in the UK atmosphere.[4,5] In view of the commercial importance of silver it is, therefore, not surprising that many commercial preparations aimed at preventing tarnishing of silver are available. These preparations fall into three categories: vapour phase inhibitors, which form an invisible protective layer on the surface of the silver; materials which absorb the corrosive gases which cause tarnishing, and polishes which deposit a tarnish inhibitor on the silver surface.

The materials tested were:

Hagerty Silver Keeper block
Shell VPI 260
3M Silver Protector strip
Hagerty Silver Keeper paper
Charcoal Cloth
Goddards Long Term Silver Cloth
Goddards Long Term Silver Polish
Goddards Long Term Silver Foam
Hagerty Long Term Silver Foam

Hagerty Silver Keeper block is a preparation containing a vapour phase inhibitor and Shell VPI 260 is dicyclohexyl ammonium nitrite, a vapour phase inhibitor for silver.[6] 3M Silver Protector strip, Hagerty Silver Keeper paper and charcoal cloth all absorb hydrogen sulphide. The long term materials are all polishes containing a tarnish inhibitor.

TEST PROCEDURE

The anti-tarnish preparations were tested by exposing them with silver test pieces to an atmosphere of hydrogen sulphide. Each product was tested in a separate quick-fit flask in which flaked rubber was used to produce the hydrogen sulphide atmosphere, which was greater than ambient concentration. The test procedure was as follows: squares of silver foil, manufactured for assay purposes, were prepared by cleaning with a glass bristle brush, degreasing in acetone and drying at room temperature. Three grams of flaked rubber was weighed into each quick-fit flask and 0.5 ml of water added to the flask. The anti-tarnish papers and cloth were cut and arranged to cover about a quarter of the internal surface of the flasks. The Silver Keeper block and open sample tubes containing weighed amounts, 1.0, 0.5, 0.25g, of VPI 260 were put into other flasks. The prepared silver pieces were suspended on nylon thread from the top of each flask. The silver polishes were tested by applying them to silver samples prepared as described above, again suspended over 3g of flaked rubber and water in a quick-fit flask. The time taken for the silver pieces to start to tarnish was noted and compared with the rate of tarnishing of a control silver sample suspended over flaked rubber.

RESULTS AND DISCUSSION

The assessment of how effective the anti-tarnish preparations were in preventing the formation of silver tarnish was made by comparing the time taken for the exposed silver pieces to show visible signs of tarnishing (Table 1).

Anti-tarnish preparations	Time for formation of a visible amount of silver sulphide (days)
Control	3
Hagerty Silver Keeper block	5
VPI 260 (1g)	5
VPI 260 (0.5g)	5
VPI 260 (0.25g)	5
3M Silver Protector Strip	11
Hagerty Silver Keeper paper	5
Charcoal Cloth	10
Goddards Long Term Silver Cloth	6
Goddards Long Term Silver Polish	4
Goddards Long Term Silver Foam	4
Hagerty Silver Spray	4

Table 1 Time taken for test pieces to tarnish

All the preparations inhibited the formation of silver sulphide on the silver test pieces when compared with the control. The greatest protection against tarnishing was given by 3M Silver Protector Strip and Charcoal Cloth. These materials absorbed hydrogen sulphide and were successful in preventing hydrogen sulphide reaching and reacting with the silver surface. The vapour phase inhibitors, Hagerty Silver Keeper and different weights of VPI 260, afforded equal protection to the silver test pieces but were not as effective as the materials which absorbed hydrogen sulphide. Goddards Long Term Silver Cloth prevented tarnishing for a longer period than the other polishes tested. In duplicate tests the order of the results reported above was reproduced, but the number of days elapsing before tarnishing became visible was slightly different. This difference appeared to be related to the ambient temperature during the tests.

All the anti-tarnish preparations become exhausted with time. When used in normal ambient conditions the effective lifetime depends on the concentration of corrosive vapour in the air. The only sign that the preparations are exhausted will be when the silver starts to tarnish, but this situation could be avoided by routinely replacing the materials at, for instance, intervals of six months.

In showcases, absorbent materials could be interleaved between backing boards and display fabrics. Replacement of the materials would require a refurbishment of the interior of the cases. In storage areas the materials could be used to line boxes and shelves. However, silver should not be allowed to come into direct contact with the charcoal cloth: in an accelerated test to evaluate corrosive effects on silver, the cloth was found to be loaded with non-volatile chlorides which corroded the silver.

Vapour phase inhibitors can be inconspicuously positioned in a showcase and are, therefore, easily replaced. The effectiveness of the vapour phase inhibitor may be reduced by its rapid loss in showcases which are not airtight. Because storage in museums tends to be in unsealed rooms in which staff often work and handle objects almost daily, the health and safety aspects of these materials should be considered before they are used.

Silver is normally polished before being put on exhibition. If a long-term polish is used, the number of times repolishing is necessary may be reduced.

CONCLUSIONS

All the products tested were to some extent effective in preventing the formation of silver sulphide on silver. The choice of an anti-tarnish preparation for use in a museum depends on the manner in which silver objects are stored and displayed. As silver objects are normally polished during conservation, a polish containing a tarnish inhibitor can be used without disruption to the normal museum routine.

REFERENCES

1. Oddy, W.A., An unsuspected danger in display, Museums Journal, 73, No. 1 (1973), pp. 27-28.

2. Blackshaw, S.M. and Daniels, V.D., Selecting safe materials for the display and storage of antiquities, in ICOM Committee for Conservation 5th Triennial Meeting, Zagreb (1978).

3. Daniels, V.D. and Blackshaw, S.M., The testing of materials for use in storage and display in museums, The Conservator, 3 (1979), pp. 16-19.

4. Graedel, T.E., Hammlott, G.W. and Franey, J.P., Carbonyl sulphide: potential agent of atmospheric corrosion, Science, 212 (1981), pp. 663-665.

5. Graedel, T.E., Franey, J.P. and Kammlott, G.W., The corrosion of copper by atmospheric sulphurous gases, Corrosion Science, 23 (11) (1983), pp. 1141-1152.

6. Shell Vapour Phase Inhibitors - VPI Range, Technical Bulletin, FC:70:55:TB, Shell Chemicals (1978).

SUPPLIERS

Hagerty Products:

W J Hagerty & Sons Ltd Inc
FL-9490 Vaduz
USA

Goddards Products:

J Goddard & Sons Ltd
Frimley Green
Surrey
England

3M Silver Protector Strip

3M
St Pauls
MN 5514
USA

VPI 260

BDH Chemicals Ltd
Broom Road
Poole
BH12 4NN
England

Tel: 0202 745520

CORROSION INHIBITORS IN ACID STRIPPING OF IRON

P.T. Wilthew

It has been proposed that in certain situations inorganic acid solutions containing organic corrosion inhibitors could be used in conservation to remove corrosion products from iron objects. Such solutions, often based on hydrochloric acid or sulphuric acid, are commonly used in industry for removing scale from iron.

Several mechanisms have been proposed for the removal of iron corrosion products by non-oxidising acid solutions, but they all require the production of hydrogen molecules or radicals by corrosion of the base metal according to the following reactions:

(1) $Fe \rightarrow Fe^{2+} + 2e^-$

(2) $2H^+ + 2e^- \rightarrow 2H°$

(3) $2H° \rightarrow H_2$

It is likely that iron oxides are first reduced and then dissolved,[1] and also physically forced away from the iron surface during the formation of bubbles of hydrogen.

The addition of inhibitors reduces the rate of either reaction (1) or reaction (2) or both, which results in a lower rate of corrosion of the base metal and a lower rate of formation of hydrogen radicals and molecules, so the rate of corrosion product removal will also be reduced. A balance must, therefore, be achieved between reducing the corrosion rate during treatment and increasing the treatment time so that the loss of base metal during the treatment is minimised.

A wide range of organic compounds has been found to inhibit iron corrosion in acidic solutions.[2,3] These include, in the case of sulphuric acid (H_2SO_4) solutions, compounds based on benzotriazole (BTA)[4,5,6] and thiourea.[6,7,8] However, the presence of additional ions, the most important of which being the chloride ion, may alter the effectiveness of the inhibitor in a sulphuric acid solution and the corrosion rate of the iron may either increase or decrease.

In the present work the amount of base metal lost during the removal of corrosion products from mild steel using 1M sulphuric acid solutions inhibited with BTA and thiourea at various concentrations was investigated. These solutions are compared to 0.5M, 1M and 2M citric acid and orthophosphoric acid (H_3PO_4) solutions. The effect of chloride ion concentration on the corrosion rate of low carbon steel in 1M sulphuric acid inhibited with BTA at a concentration of 1.10^{-2}M was also studied.

METHOD

All the sulphuric acid and orthophosphoric acid solutions were prepared by diluting the concentrated acid with distilled water and dissolving the appropriate amount of inhibitor, if any, in the dilute acid. The citric acid solutions were prepared by dissolving the solid acid in distilled water. Sufficient sodium chloride to give the required chloride ion concentration was dissolved in the acid solutions required to contain chloride ions. All tests were carried out at 291K (18°C).

The loss of base metal occurring during removal of corrosion products using acid solutions was studied by immersing samples cut from a mild steel chain in each solution. Hydrogen was bubbled through the solution for at least 24 hours before each test. The hydrogen evolved during each test was collected above the solution in a burette and the total volume evolved before the corrosion products were completely removed was determined in each case. The time at which all the corrosion products had been removed was determined visually.

The tests on the effect of chloride ion concentration on the corrosion rate of low carbon steel in 1M sulphuric acid inhibited with BTA at a concentration of 10^{-2}M were carried out using a similar procedure, but the samples used were uncorroded low carbon steel nails.

The results were converted to the weight loss values quoted below by assuming that all the hydrogen evolved was produced by reactions (2) and (3) above. No attempt was made to allow for loss of hydrogen due to the solubility of hydrogen in the base metal or reduction of the corrosion products.

RESULTS AND DISCUSSION

The results of the tests on the loss of base metal occurring during the removal of corrosion products from mild steel are given in Table 1.

Acid		Inhibitor	W(gm^{-2})	T(Hr)$\pm$10
1M H_2SO_4		-	490	20
		1.10^{-3}M BTA	445	40
		3.10^{-3}M BTA	157	60
		1.10^{-2}M BTA	36	70
		3.10^{-2}M BTA	15	60
		1.10^{-3}M Thiourea	224	50
		3.10^{-3}M Thiourea	30	50
		1.10^{-2}M Thiourea	24	60
		3.10^{-2}M Thiourea	26	50
		1.10^{-1}M Thiourea	128	40
CITRIC ACID	- 0.5M	-	37	100$\pm$20
	- 1M	-	99	90$\pm$20
	- 2M	-	94	70
H_3PO_4	- 0.5M	-	34	40
	- 1M	-	48	30
	- 2M	-	251	15$\pm$5

Table 1 Loss of base metal during removal of corrosion products using citric acid, orthophosphoric acid and inhibited sulphuric acid solutions.

W = weight loss per unit area occurring before the corrosion products were completely removed

t = time required to remove all the corrosion products

The use of 1M sulphuric acid to remove corrosion products from mild steel without a sufficient concentration of a corrosion inhibitor results in a substantial loss of base metal, and too great a concentration of thiourea has a similar result. The presence of BTA (1.10^{-2}M - 3.10^{-2}M) or thiourea (3.10^{-3}M - 3.10^{-2}M) reduces the loss of base metal which occurs to a level comparable with that obtained by using citric acid or orthophosphoric acid solutions of the concentrations studied.

The solution which caused least corrosion was 1M sulphuric acid inhibited with BTA at a concentration of 3.10^{-2}M, although low weight losses were also obtained using 1M sulphuric acid inhibited with BTA (1.10^{-2}M) or thiourea (3.10^{-3}M - 3.10^{-2}M), 0.5M citric acid and 0.5M orthophosphoric acid.

Under certain conditions, 1M sulphuric acid inhibited with benzotriazole at a concentration of $1.10^{-2}M - 3.10^{-3}M$ could be suitable for removing corrosion products from iron objects. However, the presence of chloride ions, which are often present in the corrosion products of iron objects, can affect the behaviour of inhibitors. The results of the tests to determine the effect of chloride ion concentration on the corrosion rate of low carbon steel in 1M sulphuric acid inhibited with BTA are given in Table 2, in which the percentage inhibition was defined as follows:

$$\% \text{ Inhibition} = \frac{(Ro - Rs)}{Ro} \times 100$$

Rs = corrosion rate in the test solution
Ro = corrosion rate in 1M sulphuric acid with no inhibitor or chloride ions

BTA Conc.	Cl^- Conc.	Corrosion Rate ($mgm^{-2}s^{-1}$)	% Inhibition
0	0	7.55	0
	$1.10^{-4}M$	7.72	-2
	$1.10^{-3}M$	6.28	17
	$1.10^{-2}M$	5.31	30
	$1.10^{-1}M$	3.28	57
$1.10^{-2}M$	0	0.58	92
	$1.10^{-4}M$	0.58	92
	$3.10^{-4}M$	0.50	93
	$1.10^{-3}M$	0.37	95
	$3.10^{-3}M$	0.26	97
	$1.10^{-2}M$	0.25	97
	$3.10^{-2}M$	0.23	97
	$1.10^{-1}M$	0.17	98

Table 2 Effect of chloride concentration on the corrosion rate of low carbon steel in 1M H_2SO_4 and 1M H_2SO_4 inhibited with $1.10^{-2}M$ BTA.

The presence of chloride ions did not reduce the effectiveness of BTA as a corrosion inhibitor. The corrosion rate of the low carbon steel decreased as the concentration of chloride ions increased. 1M sulphuric acid containing BTA at a concentration of between $1.10^{-2}M$ and $3.10^{-2}M$ should, therefore, be suitable in certain situations for removing corrosion products from iron objects, even in the likely presence of chloride ions in the corrosion products.

The behaviour of corrosion inhibitors in acidic solutions can be affected by several factors, including metal composition, acid concentration, inhibitor concentration and temperature, and if any of these factors are altered the effectiveness of the inhibitor should be confirmed. It would be expected that wrought iron and carbon steels would behave similarly to the low carbon steels used in this work with respect to inhibited acid stripping solutions, unless a high concentration of an impurity is present. However, if an inhibitor is not effective in a particular case, rapid corrosion could occur and an object should be carefully observed during treatment, particularly in the early stages.

Acid solutions remove corrosion products at the base metal surface and not by gradual dissolution from the surface of the corrosion product. They should only be used on objects with a substantial metal core and if complete removal of the corrosion products and slight etching of the base metal is acceptable. Inhibited inorganic acid solutions should not be used on composite or inlaid objects and thorough washing of the object after treatment to remove any traces of acid is essential.

CONCLUSIONS

The use of 1M sulphuric acid containing benzotriazole at a concentration of $1.10^{-2}M - 3.10^{-2}M$ to remove corrosion products from mild steel results, under certain conditions, in a smaller loss of base metal than the use of 0.5M - 2M citric acid or orthophosphoric acid solutions. Using 1M sulphuric acid containing thiourea at a concentration of $3.10^{-3}M$ 0 $3.10^{-2}M$ results, in certain situations, in a similar loss of base metal to that resulting from using 0.5M citric acid or orthophosphoric acid, and a lower loss of base metal than the use of 1M or 2M citric acid or orthophosphoric acid.

The presence of chloride ions did not decrease the effectiveness of benzotriazole as an inhibitor of the corrosion of low carbon steel in 1M sulphuric acid.

1M sulphuric acid solutions containing benzotriazole or thiourea as a corrosion inhibitor provide an alternative method of removing corrosion products from iron objects which could be used in conservation in situations where an acid stripping treatment is appropriate. However, careful observation is required to ensure that excessive corrosion of the base metal is not occurring unexpectedly in any area of an object, and thorough washing of the object must be carried out after treatment, as residual acid could cause serious localised corrosion later.

ACKNOWLEDGEMENTS

This work was carried out at the British Museum, London. I would like to thank Dr. V.D. Daniels of the British Museum for his help.

REFERENCES

1. Evans, U.R., The corrosion and oxidation of metals, Edward Arnold (1960).

2. Eldredge, G.G. and Warner, J.C., Unlig, H.H. (ed) Corrosion handbook, Wiley, N.Y. (1948), p. 905.

3. Trabanelli, G.T. and Carassiti, V., Mechanism and phenomology of organic inhibitors, in Fontana, M.G. and Staehle, R.W. (eds), Advances in corrosion science and technology, vol. I, Plenum Press (1970).

4. Eldakar, N. and Nobe, K., Effect of tolytriazole on iron corrosion and the hydrogen evolution in H_2SO_4, Corrosion, 32 (6) (1976), p. 230.

5. Eldaker, N. and Nobe, K., Electrochemical and corrosion behaviour of iron in the presence of substituted benzotriazoles, Corrosion, 33 (4) (1977), p. 128.

6. Wilthew, P.T., Inhibited H_2SO_4 and HCl stripping solutions for iron, ICOM Committee for Conservation, 6th Triennial Meeting, Ottawa, 81/23/6 (1981).

7. Desai, M.N., Thanki, G.H. and Gandhi, M.H., Thiourea and its derivatives as corrosion inhibitors, Anti-Corrosion (1963), p. 12.

8. Ateyo, B.G., Abo-Elkhair, B.M. and Abdel-Hamib, I.A., Thiosemicarbozide as an inhibitor for the acid corrosion of iron, Corrosion Science, 16 (1976), p. 163.

BRINCH MADSEN/BRADLEY/WILTHEW DISCUSSION

Andrew Oddy The problem is we don't quite know where we stand with regard to the law. It's one thing describing a product and testing it on its own, but it's quite another when you take a number of products and try and put them into an order of preference. As soon as you do this the manufacturer of the product which came out worst will start to say that if you had done the tests a different way their product would have come out top. I'd like to ask Susan Bradley how reproducible she thinks her experiments were?

Susan Bradley Some of the experiments were first done in 1977 and produced the same results on new samples recently, so it seems they are reasonably reproducible, but this kind of test only gives you a rough ranking. For instance, all the vapour phase inhibitors came out the same and all the absorbent cloths came out the same, but it depends on the circumstances of use which type of product you would choose.

Jonathan Ashley-Smith Goddards will give information to anyone about the composition of their products but you are not supposed to tell anyone else. There was some work at the BM some years ago on the effects of light on tarnishing rates. Were you able to confirm this?

Susan Bradley We think that there is a relationship between light and tarnishing but we haven't done enough tests to pin it down. There is less observed tarnish on samples exposed to sulphide in the absence of light but I haven't measured the thickness of the silver sulphide films.

Jonathan Ashley-Smith It has been remarked that if you use the silver cloth, which you said was very good, and then lacquer the silver, the lacquer (Frigilene) doesn't last very well. Is that correct?

Susan Bradley I have not investigated this because we have given up lacquering our silver, but it would be worth checking.

Hannah Lane Referring to the benzotriazole paper, we have often used benzotriazole dissolved in water instead of alcohol and we frequently notice in these cases growths of a hard, dark green bubble over the chloride spot. In fact, if you leave it in water with benzotriazole you can get very large growths. When this was looked at it was a complex of benzotriazole, but the scientists couldn't get any further than that. You get it within 24 hours in a water solution. It sounds like what you are getting when you expose treated bronzes to high RH.

Helge Brinch Madsen Thank you.

Julia Fenn Mrs. Bradley, have you noticed an accelerated corrosion rate after the initial protection given by any of the commercial products you have used?

Susan Bradley No, not specifically.

Janey Cronyn I would like to ask about the role of benzotriazole in Incralac. Is this an important factor? How long does it stay in the Incralac?

Helge Brinch Madsen If you treat your objects with 3% benzotriazole as I have suggested, I believe that you have a reservoir of benzotriazole left on the object and, as far as I could see from my survey of the objects, it doesn't matter whether you use Incralac or Bedacryl. But, of course, the more inhibitor you have, the better, I think. Another reason why benzotriazole has been added to Incralac is to act as an ultraviolet absorber, so in a way it is stabilising the lacquer as well.

Janey Cronyn Can the benzotriazole in the lacquer actually do anything? Is it not immobilised in the resin?

Stephen Turgoose In that context, there are plenty of examples of very much larger molecules moving through paint films; in alkyds and linseed oil based films, large molecules such as oleates and other carboxylates seem quite mobile. Very large molecules can move quite happily. I don't think there's any problem about benzotriazole.

PROTECTIVE TREATMENTS IN INDUSTRIAL ARCHAEOLOGY

R.E. Child

The conservation of industrial material (in which category is included commercial and agricultural machinery, etc) differs from traditional methods of conserving museum-type items because of the size, weight and complexity of construction of these objects. Since most of this type of material post-dates the Industrial Revolution, modern alloyds, different coating systems and advanced methods of manufacture are encountered, bringing concomitant problems of novel corrosion pathways and new degradation patterns. Furthermore, because of the size and complexity, the costs of treatments and storage and additional manpower, both for conservation and for future maintenance, become major factors in determining the style and extent of possible treatment. This is especially true if an item requires to be 'made safe' or is required 'in working order', as substantial remedial work that cannot be defined as conservation may be involved if gross replacement, alteration or addition of material is necessary.

At any stage of the conservation procedure prevention of further degradation is a priority, the use of surface protective treatments being of the utmost importance.

The following sections are designed to aid the conservator in formulating an approach to the preservation of large-scale ferrous-based objects by providing a basic guide to the steps involved between accepting a specimen and its full-scale conservation.

THE CONDITION RECORD

THE DEGREE OF RUSTING

This is usually described subjectively but is better done by direct comparison with a series of standard photographs, as in:

- BS 3900 H3 (ISO 4628/3) 'Designation of Degree of Rusting'[1] which gives a series of five black and white photographs for determining the degree of rust on coated steel.

- The European Scale of Degree of Rusting for anti-corrosive Paints[2] which gives a series of ten graded black and white photographs from Re0 = rustfree to Re9 = the whole surface rusty.

THE COLOUR OF PAINTWORK

This area of recording is often neglected owing to the difficulty of making good colour comparisons. In order to obtain accurate and representative results it is essential that the lighting is suitable, both in illuminance and in colour rendering[3] (usually daylight at 1000 to 4000 lux)[4] and that the colour to be recorded is representative of the paintwork as a whole. There are three principal systems for recording colours:[5]

- Colour photography, using a colour test chart such as Kodak Colour Control Patches or the Munsell ColorChecker Colour Rendition Chart.

- BS 5252 'Framework for Colour Co-ordination for Building Purposes' which divides colours according to their HUE (redness, yellowness or blueness, etc), GREYNESS (the amount of grey) and WEIGHT (lightness to darkness). Available not only as a card of 237 colours but also as a fan (B525 F).

- Munsell Book of Colour - a file of 1,488 colours arranged according to their HUE (redness, yellowness, etc), VALUE (lightness or darkness) and CHROMA (greyness). It is available in a variety of forms, including a matt and gloss version.

As colours may match under some lights but not under others, owing to pigmentation differences - a phenomenon known as 'metamerism' - it is recommended that colours be compared both under good diffuse daylight and under tungsten light.

THE TEXTURE OF SURFACES[6]

This is worth noting briefly even though an object's present appearance is probably due to weathering, etc, not the originally applied finish. A simple convenient method divides the definitions into: full gloss finish, eggshell (semi-gloss) finish, silk (semi-gloss) finish, matt finish.[7]

SURFACE CLEANING

NON-DESTRUCTIVE CLEANING[8]

Used to remove masking layers of mud or other caked deposits that might act as a moisture trap, soluble salts that might encourage electrolytic corrosion (salt, sweat, acid rain, dilute inorganic inhibitors) and greasy/oily deposits that can stain paintwork and affect the adhesion of future surface treatments. The cleaning process should not materially affect the object being cleaned but expose the whole for better examination and treatment. The principal methods are:

- Brush or spray application of the cleaning preparation followed by water-hose rinsing and air drying. The types of cleaner are in Table 1:

Material	Removes	Advantages	Do not use on
Emulsifiable solvent systems (white spirit & emulsifier)	Thick oil grease and dirt	Can be used on metals, most paints	Organic materials - textiles, rubber
Aqueous alkaline/ detergent (sodium hydroxide type)	Oil, grease and dirt	Effective on bare ferrous metals	Aluminium, tinplate, galvanised iron, organics
Sugar soap	Light coatings of oil, grease, dirt	Inexpensive easy to use	Some paint films softened
Detergents	Light coatings of oil, grease, dirt	Effective on most surfaces	Some paint films or where frothing is a problem
Organic solvents (paraffin, white spirit)	Oil, grease but not dirt efficiently	Good degreasing action but may smear oil over wider area	Organic materials or areas difficult to rinse

Table 1 Non-destructive cleaning methods

- High pressure hot water cleaners, with the facility to feed in additives to the water flow, perform the same function as brush or spray applications but faster and to a higher standard of cleanliness. A suitable degreasant is paraffin fed into the water stream at 2% followed by copious rinsing with hot water and drying.

- Steam cleaning removes oil and grease residues readily owing to its high temperature, but lacks the 'punch' of high-pressure hot water cleaning and the rinsing powers of water spraying methods, which remove the soluble surface contaminants.

DESTRUCTIVE CLEANING METHODS[9]

These methods remove not only dirt and caked on deposits but also old or degraded surface layers (paints, varnishes, rust, etc). When they can ethically be used they perform the function of cleaning, degreasing, paint stripping and derusting in a rapid and efficient manner. The principal methods are summarised in Table 2:

Method	Advantages	Disadvantages
Wire-brushing (hand or powered)	Cheap, easy, safe, effective on light rust	Only removes loose rust; burnishes underlying rust
Sand blasting (dry)	Effective, no moisture, fine abrasive removes some surface contaminants	Expensive, dusty and unpleasant for operatives
Sand blasting (wet)	Effective, removes soluble contaminants, no dust	Drying problems cause rerusting
Flame cleaning (oxyacetylene flame)	Removes moisture burns off part scale, etc	Buckles thin metal. Not very effective
Chemical methods (acid pickling)	Effective, controllable, can leave inhibited surface	Dangerous; needs thorough rinsing and drying, attacks tin, zinc coatings

Table 2 Destructive cleaning methods

SURFACE PROTECTION

The application of protective coatings serves two main functions in the prevention of corrosion: the exclusion of oxygen, water and contaminants and the chemical inhibition of corrosive attack. The effectiveness of any surface coating in protecting its substrate depends on many factors, especially:

- Cleanliness of substrate of oil, grease, rust and electrolytes;

- Quality of surface to be coated;[10]

- Method of application of coating (brush, roller, spray).[11] Normally brush application is to be preferred, at least for the initial coat, to achieve good adhesion with the substrate.

TRANSPARENT REMOVABLE COATINGS

These are summarised in Table 3:

Coating	Advantages	Disadvantages
Silicone sprays liquids (eg WD 40) Silicone waxes (eg Waxoyl)	Water displacing, harmless to most surfaces; easy to apply and remove	Leaves grease film, impermanent needs to be frequently replaced. May stain surrounding material
Mineral oils and waxes (may be specialised with inhibitors, etc)	May displace water; effective in thick coatings, easily removable	Oily film, may slowly oxidise and go gummy; may stain surrounding material
Varnishes and lacquers (eg polyurethanes, yacht varnish)	Cheap, easy to apply	May yellow, become non-reversible craze, permeable

Table 3 Transparent removable coatings

PIGMENTED COATINGS (usually non-reversible practicably)

These are usually paints or pigmented lacquers, varnishes and resins and provide the most practical coatings for the prevention of corrosion. Their use is restricted to those applications where an opaque non-reversible coating is acceptable - for instance, where there is no feasible alternative for the long-term preservation of an iron object where the surface appearance is not of interest or historical value. There are numerous types of paint systems marketed for their rust-preventing capabilities; red lead metal primer is considered to be the most efficient primer for bare ironwork, but it depends on a linseed oil medium which may give a poor finish and be incompatible with top coats and, owing to its toxicity, it has largely been replaced by zinc phosphate or calcium plumbate primers. Undercoats and topcoats must be compatible with the primer, otherwise painting defects result. For severe environment special coatings may be chosen, such as epoxy resin paints, micaceous iron oxide paints and bituminous paints.[12]

ENVIRONMENTAL CONTROL

The general principles of providing a controlled environment for the benefit of specimens and personnel are well known. However, these principles are, in general, applied to environments inside well-constructed and maintained buildings which can be heated, lit and cleaned. Large industrial items, however, may have to be stored or displayed in the open or in covered sheds at the mercy of wind-borne pollutants and the extremes of external temperatures and humidities. In accepting storage conditions of this type it may be useful to bear in mind the following:

SITING OF STORES

In order to determine the aggressiveness of the conditions the contents will contend with, the following must be considered:

- Degree of exposure,[13] with inland areas classed as 'moderate', to coastal areas, especially on the west coast, classed up to 'severe'. Nearby shelter from trees, other buildings, etc, may moderate conditions while high-rise structures, exposed sites, etc, cause more severe conditions.

- Wind-blown contaminants and pollutants may include not only rain but also smoke, ash, fumes, etc, from nearby industry; salt-laden spray from coastal areas which can penetrate 8km inland.

- Local wind phenomena causing dust, dead leaves, etc, to collect in the storage area.

LOCAL ENVIRONMENTAL CONDITIONS

Meteorological offices give local information on temperature, humidities, wind direction and speed. From this information hazards such as high ambient relative humidities and low temperatures can be assessed. These may mean that there is a potential danger of condensation if the stored material comes to be at a lower temperature than the surrounding air.[14] Consider the following case:

Average January temperature in Cardiff is 5°C
" " R.H. " " " 80-90%

From hygrometric tables:[15]

Average January Dew Point in Cardiff is 3°C

If an object in an unheated store is at a lower temperature than 5°C, which can easily happen when the building is warming up after a cold night, condensation will occur on the object until it warms up to 3°C. This is avoidable with well-ventilated, heated stores or if dehumidifiers or desiccants are used.[16]

HANDLING

This is probably the greatest cause of damage to large specimens in museums, even though the precautions that should be taken are well known. The following points are worth remembering:

- gloves should always be worn during manual handling to avoid contamination with sweat;[17]

- objects should always be raised off the floor on wood blocks to aid handling and to avoid damp, etc, from floors;

- after any moving operation the object's surface finish should be inspected and repaired as necessary.

REFERENCES

1. British Standard BS 3900: Part H3 1983 (ISO 4628/3 1982) British Standard Methods of tests for paints, British Standard Institute, London.

2. European Committee of Paint, Printing Ink and Artists' Colours Manufacturers Associations, The European scale of degree of rusting for anti-corrosive paints.

3. Thompson, G., The museum environment, Butterworths (1978).

4. British Standard BS 3900: Part DI 1978, Visual comparison of the colour of paints, British Standard Institute, London.

5. Chamberlin, G.J. and D.G., Colour: its measurement, computation and application, Heyden Press (1981).

6. British Standard BS 4800: 1981, Paint colours for building purposes, British Standard Institute, London.

7. Brill, T.B., Light, its interaction with art and antiquities, Plenum Press (1980), p. 99.

8. British Standard BS 1133: Section 6: 1966, Packaging code, British Standard Institute, London.

9. British Standard BS 5493: 1977, Protective coating of iron and steel structures against corrosion, British Standard Institute, London.

10. Department of Industry, Guides to practice in corrosion control no. 13: Surface preparation for painting, H.M.S.O. (1982).

11. Department of Industry, Guides to practice in corrosion control no. 12: Paint for the protection of structural steelwork, H.M.S.O. (1982).

12. Fulcher, Rhodes, Stewart, Tickle and Windsor, Painting and decorating: an information manual, Granada Building Crafts Series (1981).

13. British Standard BS 5502: Section 3.2:1978, Design of buildings and structures for agriculture, British Standard Institute, London.

14. Paterson, M.P., Aeroplanes in museums - creating an atmosphere, Preservation of technological material, proceedings of symposium in the Science Museum, 10th May 1974, Science Museum (1976), London.

15. H.M.S.O., Hygrometric Tables Part III, Aspirated psychrometer readings degrees celsius, H.M.S.O. (1964).

16. British Standard BS 1133: Section 19: 1968, Packaging code - use of desiccants in packaging, British Standard Institute, London.

17. Lind, S.E., Corrosion of metals by human sweat and its prevention, Corrosion Science, 12 (1972), pp. 749-755.

SURFACE TREATMENTS FOR LOCAL HISTORY COLLECTIONS

Sylvia J. Turner

Norfolk Museums Service has extensive collections of metalwork of marine, agricultural and domestic origin. Whilst there is much published on the treatment of archaeological ironwork, there is very little written in the conservation journals on the treatment of historical material. In the past ten years we have used several different treatments and tested others.

CRITERIA FOR SELECTING TREATMENTS

- Type of finish required: is there residual paint which must be retained; is the colour of finish important; is a matt or shiny finish equally acceptable?
- Quality of finish: does the finish give a hard surface or does it dry to a tacky finish that will mark visitors' clothing, or in time attract dust and look unsightly?
- Durability of finish and use of object: is the object intended for storage or display; is it likely to be moved frequently causing damage to the coating; will it be protected by a showcase or open display?
- Complexity of the object: is it one part with a simple shape or many delicate intricate parts, perhaps including other metals or non-metals?
- Reversibility: is the finish readily removable when required?
- Health and safety: does the finish have any harmful effects?
- Cost effectiveness: when large numbers of objects are to be treated, protective coatings that are both effective and also economical in purchase cost and labour are important.

TREATMENTS

TANNIN BASED

Tannin treatments have been used in industry for many years and the action of tannin as an antioxidant capable of readily forming iron tannate complexes is known. The iron tannates form a dense black insoluble protective layer over the metal surface. The basic treatment of mechanical cleaning followed by the application of a thin tannin wash was used when treating the cast-iron roof over 'Big Ben' in the 1950's.[1] Because the reaction is oxygen-dependent it is important to apply thin washes, allowing the thickness of coating to gradually build up. It is possible to apply paint over tannin-coated surfaces and some physical barrier is necessary, as the tannate layer will rub off.

The method of Pelikan[2] was the one which we used from 1975 for treating objects like mantraps. Objects so treated that have been on show ever since in the Castle dungeons show no sign of re-corrosion. Several points are worth mentioning. Firstly, the preparation of the surface is most important: failure to remove residual grease will prevent the formation of the tannate layer. The treatment is really only suitable for lightly corroded objects where the rust will be in immediate contact with the applied tannin. The application of several coats followed by a protective lacquer - we used Ercalene (cellulose lacquer) - makes this a lengthy process. Care should be taken when tannic acid is used, as it is a suspect carcinogen.[3]

It is now possible to buy commercial formulations of tannin, such as Trustan 23, though this is comparatively expensive for large-scale use. A similar product, Fertan, has been used recently on the Holland I submarine. Fertan has been tested at the British Museum, where it was noted that it produced a shiny black surface layer but underneath the rust was untouched.

RUST REMOVERS

The major product of this type that we have tried is Jenolite RRPL. Earlier de-rusting treatments in large tanks proved too time-consuming because of the large number of composite objects in the collection. Following the removal of grease and loose rust, the Jenolite is applied with a brush and, after 15 minutes, the excess is removed with a cloth. There is no need for any subsequent rinsing. The rust is removed, leaving the surface of the object coated with an inert layer of iron phosphate.

Jenolite is available in two thickened and jelly forms which enable the process to be carried out without immersion. Whilst the Jenolite itself will not attack original paint, the formation of the iron phosphate layer gives the object a grey appearance which we found to be unacceptable and is really only suitable if the object is going to be repainted. As an acid treatment, it does carry some risk to the operator.

TEMPORARY PROTECTIVES

There are now a number of products made to protect metal surfaces during transportation and in storage. Their main characteristic is that they can readily be removed by common solvents. They act by forming a moisture impervious barrier or by a process of chemical inhibition, so that the rate of corrosion caused by the presence of moisture is substantially reduced. In 1974 the use of Ensis Fluid grade 256 was recommended by the Museum of English Rural Life, Reading. Grade 256 (now SDC) is a de-watering fluid produced by Shell which gives a soft thick film, providing indoor protection for approximately 18 months. Visually, Ensis is pale amber and transparent. Its greatest disadvantage is the softness of the finish: it rapidly picks up dust and becomes unsightly. The film is also likely to be damaged by handling. Another de-watering anti-rust product is Dispen which leaves a light lubrication film.

Waxoyl and Weather-X are two similar products. Waxoyl consists of wax particles and a viscous inhibitor fluid in suspension in white spirit. After evaporation of the solvent, the surface is dry to the touch, but it is claimed to have "self-healing" properties which allow the natural repair of small physical disruptions of the surface. This is significant, for museum objects may well be handled or moved about. In fact, Waxoyl is specifically recommended for situations where both a lubricant and corrosion preventer are required. In appearance it is brown and semi-transparent; this is not unpleasant. It is claimed that after 48 hours drying it is possible to remove the wax, leaving the inhibited surface which can then be over-painted. Unless repainting is intended, it is not necessary to prepare a clean metal surface before applying the Waxoyl. Such cleaning as is done need only be sufficiently thorough to make the object visually acceptable for display.

Weather-X is, again, a wax with an inhibitor. The film, however, tends to be thicker and more visible.

LACQUERS

The final products I have listed are ones that I have tested in the past but have not had occasion to use. They are:

Black Seelfos: After the removal of rust and grease the Seelfos is applied by brush or immersion. A thin layer of iron phosphate is formed, sealed by a thin coating of resin. As its name implies, the lacquer is pigmented.

Superaquapel RF: A solution in which the object is placed after, say, patination or tannin treatment which has coloured the metal surface. All water is removed and a thin film of clear alkyd resin is deposited.

TESTS FOR FINISHES

Two tests which have proved useful in assessing the suitability of any product are tests for corrosion at high temperature and high humidity and tests for adhesion and stickiness.

TEST FOR PROTECTION AGAINST CORROSION AT HIGH TEMPERATURE AND HIGH HUMIDITY

Mild steel panels 150mm x 100mm x 1.26mm were prepared by abrading with silicon carbide paper to a 600 grit finish. They were then weighed, coated and, after 48 hours, re-weighed and placed in a 100% RH environment at a temperature of 42° - 48°C. The degree of protection was assessed on the basis of the weight of coating per panel, the time at which corrosion was initially observed, and the final appearance of the panel.

RESULTS (Table 1):

Product	Weight of coating/panel (g)	Onset of corrosion (days)	Severity of corrosion after 500 hours Visual Inspection
Weather-X	0.5	4	light
Waxoyl	0.3	4	light
Trustan	0.6	1	severe
Black Seelfos	0.6	1	moderate
Aquapel	0.1	2	light
Dispen	0.1	1	moderate

Table 1 Protection against corrosion at high temperature and humidity

Under the severe test conditions all the coatings eventually broke down. However, Weather-X and Waxoyl were more effective than other products. Of the two, Waxoyl is less conspicuous when applied.

TESTS FOR ADHESION AND STICKINESS

Mild steel panels were prepared and coated. After 24 hours they were laid horizontally and each was covered with a piece of Kraft paper, a layer of felt 10cm square and 6mm thick, and a 0.9kg weight. After one hour the weight was removed and the surface of the panels examined for exposure of the metal.[4]

RESULTS (Table 2):

Product	Loss of coating
Tannic acid (Pelikan)	no loss
Tannic acid (11) + Ercalene	no loss
Trustan	small specks lost
Waxoyl	no loss
Weather-X	no loss
Aquapel	no loss
Seelfos	no loss
Dispen	large loss, paper stained

Table 2 Adhesion of coatings.

Of the three tannic acid treatments tried, the Pelikan based recipes seemed to perform better than the commercial product. The soft oil film formed by Dispen was rapidly damaged by contact.

CONCLUSIONS

There are a number of products available which will give some protection to ironwork. Which one is used will depend in practice on a number of variable factors, but Waxoyl in particular seems to have many advantages.

REFERENCES

1. Oil and Colour Chemists Association, Tannins v. corrosion, Paint, Oil and Colour Journal, April 5 (1957), pp. 793-4.

2. Pelikan, J.B., Conservation of iron with tannin, Studies in Conservation, 11 (1966), pp. 109-115.

3. Clydesdale, A. (Ed), Chemicals in conservation, Conservation Bureau, Scottish Development Agency, Scottish Society for Conservation and Restoration (1982).

4. British Standard BS 1133: Section 6 (1966), Temporary protection of metal surfaces against corrosion (during transport and storage), British Standard Institute, London.

CHILD/TURNER DISCUSSION

David Scott Sylvia Turner, have you also evaluated Pelikan's polyphosphate methods, published in 'Studies'?

Sylvia Turner No.

James Argo I did look at some of Pelikan's work, specifically I was looking to see the effects of acidification with tannic acid, and I found we were getting better results with tannic acid if we acidified it. You get a better film if there is a balance of iron, water and tannate. You need a pH of around 2.5. I found phosphoric acid excellent; other acids worked.

Todor Stambolov The idea of using tannins in iron conservation was first suggested in the '60s by Mr. Leo Biek, who discusses it in his book 'Archaeology under the microscope'. It was also published in 1912 in the 'Chemical Journal'.

Wendy Robinson Does anyone know what is the optimum temperature and RH for applying lacquers, and which temperature and RH might actually be harmful?

Bob Child All we do is make sure that the temperature of the object is above the dew point, and that the cooling effect of the evaporation of solvent doesn't bring it below that point.

Bruno Mulethaler Sylvia Turner, have you studied all the products available for use in the ways you describe?

Sylvia Turner No. A large number of products had to be rejected because they did not fulfil the various criteria. I tried to find products representative of various types rather than trying all products of every type.

AMINES AND IRON STABILISATION

DISCUSSION

Andrew Oddy (Chair) At this stage Dr. Argo will describe his work on amines and there will then be a discussion including Dr. Turgoose, who has also been working in this field.

James Argo Thank you, Mr. Oddy. I will be reporting today on some of the work I have been doing with amines. In particular, I'm going to be talking about experiments which indicate that amines have been retained within the corrosion product layer. The actual benefit of this I will allude to at the end of my brief discussion. These experiments have been conducted using goethite as a substrate, by exposing the goethite to solutions containing a variety of amines, including ethylenediamine and triethanolamine.

These materials have been examined using the methods of thermal analysis, and I will briefly inform you of this technique. A sample is placed on a sensitive weighing apparatus and subjected to a linear increase in temperature. At certain temperatures a loss of mass is detected and this mass loss is associated with an energy loss. The net result of the analysis with this analytical method is to produce a table such as Table 1, where 'binding energies' for some amines on goethite are shown. Binding energy here is the magnitude of the energy associated with the presence of the amine on the surface of the corrosion product. As you see from the table, these energies are of the order of 34-35 kilocalories per mole, and this is closely comparable with chemical bond energies. This indicates that the amines have been strongly chemisorbed on the surface of the goethite and a considerable amount of energy must be expended in order to remove them. They can be considered as quite stable.

Decorator	Symbol	Conc %	Enthalpy Joules/ gm	Mass %	Binding energy KCal/mole
ethylenediamine	EN	10	92.00	3.94	33.4
ethylenediamine	EN	5	48.00	1.97	34.9
triethanolamine	TEA	1	176.00	18.18	34.5
diethylenetriamine	DIEN	2	61.40	3.95	38.2
hexyldiamine	HDA	2	3.48	0.31	31.3*
hexyldiamine	HDA	2	19.60	1.00	10.3**
pyridine	PY	2	22.70	4.16	10.3
n-butyl amine	N-BUA	2	13.10	0.98	23.3
acetyl acetone	ACAC	2	70.20	19.57	8.6

* A site ** C site

Table 1 Binding energy: selected decorators on goethite

I have found that the amines are similarly chemisorbed onto the surface of haematite and magnetite. The effect of the amine appears to be to impede the access of atmospheric water to the surface. The material may be considered as having been rendered hydrophobic by the presence of the amine; this is the direct result of the chemisorption of the amine onto the corrosion product layer.

Making the surface hydrophobic, or somewhat more hydrophobic than it was, has the effect of allowing treated objects to be exposed to a fluctuating RH, or to a high RH value, without corrosion occurring. Professor Evans from Cambridge, together with Dr. Taylor, demonstrated that there is a critical RH for iron when exposed to water in the presence of sea salt. Prof. Evans demonstrated that this critical humidity was of the order of 40%. This means that at RH values in excess of 40% rusting occurs rapidly, whereas at RH values of less than 40% rusting occurs slowly or not at all. At the critical humidity value, a film of liquid was found to form. This film of liquid forms an electrolyte which will permit an electrochemical cell to form or a cathodic reaction to begin. The net result is rapid corrosion and degradation of the object. At lower RHs corrosion is non-existent or much slower, due to the absence of an electrochemical cycle. I have found that the presence of the amine chemisorbed on the goethite layer has apparently succeeded in raising the critical humidity to a much higher value. It was found that treated samples are stable to RHs in excess of 80% for periods of six months or more.

I will briefly outline the procedure that is employed at present in CCI. This involves the use of a bath of 5% ethylenediamine in water. About 4 kg of iron is immersed in about 16l of solution. The bath is maintained at about 70°C in a fume hood. The process is continued until after about 7-10 days analysis indicates that no further chloride is being extracted in that bath. The solution is then changed and the procedure is repeated. Again, a plateau is reached and the solution is changed. This continues until successive plateaux are the same from one bath to the next, which shows that we are on the flat portion of the extraction curve. This usually occurs after about 50-60 days, depending on the number of objects being treated. The objects are removed, washed in water until a pH of 7-8 is reached, and dried through acetone. I have found that they are extremely stable; nor do they tend to react to excesses of RH. I looked at about 3,000 pieces of iron last year and there was a failure rate, defined as the appearance of new corrosion, of less than like 2%.

We get down to about 5 ppm of chloride in solution. By total analysis of the iron I have found that the initial chloride concentration averages about 9600 ppm. The final concentration of chloride in the corrosion layers may be about 150 ppm.

The chloride analysis method is to take a sample of, say, 5 ml (if we are using 5% ethylene diamine) and titrate it with 10% sulphuric acid. It is titrated to an end point of pH 5 which is a methyl red indicator. An equivalent quantity, 5 ml, of pH 5 buffer is added, and the quantity of chloride read using a specific ion electrode against a standard titration curve. It has been found necessary to fill the inner junction of the reference electrode with 5% ethylenediamine to eliminate transient and stray currents. This has given quite reproducible readings; when I compared these to atomic adsorption experiments I found that the analysis was more accurate than atomic adsorption at low values and, as would be expected, less accurate at high values.

Stephen Turgoose Have you found any relationship between the pH of the solution and the effects of the amine? The natural pH is above 10.

James Argo I have not found any evidence that pH is a major factor in this strong chemisorption.

Stephen Turgoose I have looked at some other amines and benzotriazole is clearly quite strongly adsorbed on a magnetite surface.

James Argo What pH were you looking at there?

Stephen Turgoose That was at natural pH, about 5-6. It was clearly strongly adsorbed, so in a sense there may well be an effect of cathodic inhibition on oxide cathodes.

James Argo In fact it seems at this stage that what we're getting on iron with ethylenediamine or triethanolamine is behaviour very similar to that of benzotriazole on bronzes.

Stephen Turgoose Just using the natural pH of the amine solution?

James Argo Yes, in fact we're using about a 5% solution of ethylene diamine, but I have used as low as 1% with triethanolamine and I believe Eli Anderson has been using

0.5% of triethanolamine.

Andrew Oddy What solvents were you using?

James Argo Always water.

Sarah Pollard Has the iron been cleaned beforehand, or is it as excavated?

James Argo It has usually been in some sort of a holding solution, but apart from that nothing else.

Sarah Pollard So it's got all the outer layers?

James Argo Often, yes. It's done in a batch process and it's better if you can remove the outer layers, because the first batch is very dirty.

Rose Johnson How does the amount of chloride that comes out with this technique compare with the amount of chloride that comes out with alkaline sulphite or deionised water washing?

James Argo I haven't compared it rigorously to alkaline sulphite. I think the extraction is comparable. It's very much faster than distilled water.

Andrew Oddy Does it change the surface appearance? Does it blacken the surface?

James Argo There is some formation of magnetite, yes.

Mike Corfield One of our students, Carol Browne, has used this treatment for the past year. We've treated about 500 nails, Roman and medieval, with the same sort of process but using simplified equipment, sealed plastic boxes in a water bath, and after treatment subjected the material to high humidity, 95% RH. We have found the same sort of results; again there is a low rate of breaking out; it looks very promising. It's fairly disastrous if there is any material other than iron, including modern repairs using resins and so on, or other metals, copper, silver, tin especially.

James Argo Yes, it should not be used when any materials other than iron are present. It has been used on small objects but Ian Cross of Parks Canada has also used a poultice-type method on large objects, such as anchors, which were quite successfully treated.

Suzanne Keene Can I ask Stephen Turgoose whether his results are generally in agreement with Jim Argo's?

Stephen Turgoose I would agree that a range of amines can strongly bind to iron oxide surfaces. A whole range of amines seems to adhere quite strongly, not just these aliphatic ones. My feeling is that part of the reason for the effectiveness of the washing might be that there is no precipitation of ferrous hydroxide; although you are using an alkaline solution, there is this formation of a complex with ferrous ions. So by using amines you can use an alkaline solution and have no precipitation in the pores, which you can't do by using hydroxide solutions.

James Argo I quite agree with that.

Suzanne Keene I have been using Shell VPI 260 (dicyclohexyl ammonium nitrite) in soaking iron, which does prevent the formation of the flocculant orange stuff you often get on the objects and in the water, but it certainly doesn't result in very effective chloride removal, so perhaps this is not the main factor involved.

Stephen Turgoose That you don't see precipitated hydroxides may indicate that they are being formed within the corrosion products.

Suzanne Keene So the precipitate on the objects and in the solution is not the precipitate you meant which interferes with chloride removal?

Stephen Turgoose No, the interfering precipitate is the ferrous hydroxide that can form within the object, and there are complexes of ferrous ions with many of these amines, which may prevent its formation.

James Argo I should emphasise that I was looking at these particular amines because I found that they were effective in transforming akaganeite into something else, in fact magnetite. I focused my attention on these rather than going into something else, like BTA.

Stephen Turgoose I was looking at things which adsorbed onto magnetite as cathodic inhibitors, which is a slightly different emphasis.

James Argo I believe that they are cathodic inhibitors because they prevent the water from getting in to form an electrolyte or promote the cathodic reduction of water.

Andrew Oddy I seem to remember that there are lots of health and safety questions about aromatic amines.

James Argo I didn't work on cyclohexylamine because it has too many health risks. Ethylene diamine is listed in some references as being a middle level health hazard. Triethanolamine seems to be a completely innocuous material; it has no health hazards that I could find. By the way, amines are not mentioned in the SSCR publication Chemicals in conservation.

Stephen Turgoose All the amines should be handled with care but some are really quite toxic. Quoted TLV's (Hazards in the Chemical Laboratory, ed D.G. Muir, The Chemical Society, London 1977) are: cyclohexylamine, 10 ppm (40 mg m^{-3}); ethylenediamine 10 ppm (25 mg m^{-3}).

THE COPPERGATE HELMET - A HOLDING OPERATION

James A. Spriggs

The temporary storage of composite objects from waterlogged levels has always been fraught with difficulties, and when one of the components is iron the problem is particularly acute. The chief prerequisite for a waterlogged organic material is that it be kept wet, but if the other material is iron, then the requirements of the two materials are almost mutually exclusive, since most archaeological iron will oxidise rapidly in the presence of moisture and oxygen (air). Little has been published on either successes or failures at attempts to store ironwork damp, apart from some trials with vapour phase inhibitors (eg by Spriggs[1]) or by total immersion in an alkaline solution (eg by North and Pearson[2]).

The circumstances of the discovery of the Coppergate Anglo-Saxon helmet by a Hymac grading machine, on May 12th, 1982, is amply described elsewhere.[3] It was found approximately eight metres below street level, at the bottom of a square timber-lined pit, possibly a well, cut deep into natural clay. The helmet was found surrounded by a largely organic black deposit containing many fragments of wood and twigs, and also large lumps of clay. The pit contents seem to have remained totally waterlogged since their deposition. The helmet itself, constructed from rivetted iron plates with brass fittings, was likewise full of clay and organic debris, but also contained a number of other metallic items. On discovery, the surface of the iron plates was barely affected by corrosion, with only a light covering of greyish mineralisation, which after subsequent removal revealed original surface detail to a quite remarkable degree. The brass components were similarly in a fine state of preservation, being in the most part still shiny under a thin covering of black corrosion, mud and clay. Preliminary discussion raised the suggestion that the helmet might still contain the remains of a leather or textile cap which would probably have been worn under the helmet. This possibility gave rise to the need to 'hold' the helmet, with its contents intact, until further research and discussions had taken place.

From previous experiences with well preserved ironwork from waterlogged deposits, it was clear that the helmet could not be stored damp in an oxygenated atmosphere, as rapid corrosion would undoubtedly take place. Storage by immersion in dilute alkali was not considered suitable, owing to the loose and soluble nature of the helmet contents, and our previous attempts at using VPIs had not proved them to be entirely satisfactory in damp conditions. Therefore, damp storage in an inert atmosphere seemed the only choice, although the technique had not to our knowledge been used previously for the storage of unstable ironwork. Inert atmospheres are, however, used during the preparation of mineral and metallographic samples for analysis[4] and the method has subsequently been suggested for artefacts in Turgoose.[5]

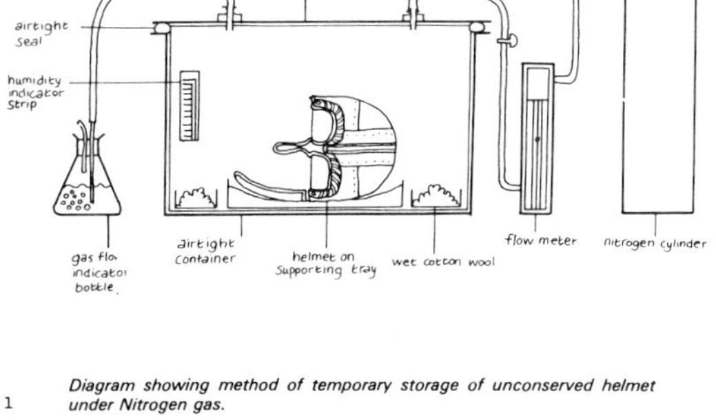

Fig 1 *Diagram showing method of temporary storage of unconserved helmet under Nitrogen gas.*

Within 24 hours of the discovery of the helmet, we were able to transfer it from the temporary packaging in which it was lifted from the ground into an inert atmosphere of nitrogen gas. Fig. 1 shows the helmet in an airtight plastic container, connected to a cylinder of oxygen-free nitrogen gas. A high relative humidity was maintained by surrounding the helmet with a number of petri dishes containing wet cotton wool. A check on the humidity was kept using RH indicator strips. The container was provided with a clear acrylic (Perspex) lid, made airtight by being bedded onto a silicone rubber seal with bolts and wing-nuts. Plastic tubing was used to connect the gas cylinder to the container via a needle valve and a flow meter, a brass gland leading the tubing through the Perspex lid (Fig. 1). The gas flow was regulated so that a slow continuous stream of bubbles issued through the gas flow indicator bottle (see diagram). It was found that a size 'J' B.O.C. nitrogen gas cylinder - about 4 ft tall - would last ten to fourteen days.[6]

The helmet remained in the nitrogen atmosphere for a period of five weeks, during which time the seal on the container was broken only four times, to permit radiography and other recording to take place. During these brief outings, no longer than two hours at the most on each occasion, some of the greyish corrosion products on the surface of the iron plates appeared to be disrupted by a bright orange/red product - a startling indication of how necessary storage in the nitrogen atmosphere was. Active corrosion, including weeping, was also noticeable around areas of damage on the helmet, which had mainly been caused by its somewhat violent mode of discovery on site. However, at no time were there visible changes while the helmet was sealed in nitrogen. The clear Perspex lid made it possible to do a certain amount of recording and examination without having to remove the helmet from its container. By disconnecting the supply and exhaust tubes and quickly sealing the connector glands in the lid with plastics putty, it was possible to transport the helmet to the York District and General Hospital and back again for radiography without having to break the seal. Although the plastic of the lid and container would easily have been penetrated by 'X' rays, for the sake of convenience most of the radiography was done with the helmet removed from the container.

During this holding operation, the programme of conservation and research work on the helmet had been decided on, and in mid-June the contents of the helmet were carefully excavated and removed.[7] Once this was completed, there was no further need to store the helmet in a damp atmosphere, so it was dried over silica gel and transferred to a specially built clear Perspex desiccator cabinet for longer term storage.[8] Since being dried out, the helmet has been removed from the desiccator on many occasions for conservation and recording, but no further signs of corrosion have been noticed.

Despite the slight disappointment of not discovering the remains of the cap or helmet lining that we had hoped for, this 'holding' operation using nitrogen gas nevertheless proved that theoretical techniques can sometimes be made to work very well in practice.

ACKNOWLEDGEMENTS

Thanks are due to staff of York University Biology Department Workshops and Stores for providing much of the equipment, and to Dr. David Scott for his advice.

REFERENCES

1. Spriggs, J.A., The recovery and storage of waterlogged materials at York, The Conservator, 4 (1980), pp. 19-24.

2. North, N.A. and Pearson, C., Investigations into conserving relics from the sea, Leigh, D. (ed) Conservation in archaeology and the applied arts, (1975

Stockholm Conference), I.I.C., London (1975), pp. 73-78.

3. Addyman, P.V., Pearson, N., Tweddle, D., The Coppergate helmet, Antiquity, LVI (1982), pp. 189-194.

4. North, N.A., Corrosion products on marine iron, Studies in Conservation, 27 (2) (1982), pp. 75-83.

5. Turgoose, S., Post-excavation changes in iron antiquities, Studies in Conservation, 27 (1982), pp. 97-101.

6. The current price for a size 'J' cylinder of 'oxygen-free' nitrogen from B.O.C. is approximately 4.00, excluding hire fee and deposit on cylinder and valve gear.

7. Spriggs, J.A., Conservation of the helmet, Interim, 8 (4) (1982), pp. 29-36.

8. Clegg, P. and O'Connor, S., Head start, Conservation News, 20 (1983).

HOLLAND 1 - AN IN-DEPTH APPRAISAL OF ITS CORROSION

M.J. Schofield and A.N. Rothwell

The Holland 1, the Royal Navy's first submarine, was recovered in August 1982 after spending 69 years on the sea bed. Built in 1901, the Holland 1 was obsolete before being put into service. It was on its way to Milford Haven to be scrapped when it sank near the Eddystone lighthouse in October 1913. Until it was recovered it lay in 200 feet of water and it was presumed by the staff of the Royal Navy Submarine Museum, Gosport that it would be badly corroded. However, after its recovery, it was reported (New Scientist 31st March, 1983, p. 905) that the Holland 1 was in good condition and had suffered very little corrosion. The Corrosion and Protection Centre Industrial Services (CAPCIS) at UMIST was very interested in this seemingly unusual case. CAPCIS is a corrosion consultancy providing expertise and testing for all types of industry, including many offshore applications. In these very special circumstances this investigation was carried out with no charge to the Royal Navy Submarine Museum. The facilities at our disposal for failure investigation have enabled corrosion scientists at CAPCIS to explain the corrosion condition of the Holland 1.

CAPCIS visited the Holland 1 when it was safely in dry dock at Devonport. It had been cleaned by high pressure water blasting and treated with a corrosion-inhibiting product. This was performed because the museum staff feared that, once exposed to the air, it would corrode rapidly or even disintegrate. Such disintegration is a well-known phenomenon but is restricted to cast iron objects which have suffered extensive graphitisation. In this corrosion mechanism the iron is slowly dissolved from the cast iron, leaving the other constituents, principally graphite. The graphite is in the form of a network, retaining the shape but not the strength of the original article. A cast iron article which has been mostly, but not completely, graphitised contains some iron in the +2 oxidation state. On exposure to air, oxidation to the familiar rust (Fe_2O_3 n H_2O) takes place. This is an exothermic reaction, the heat produced being sufficient to cause disintegration of the relic. Canon balls and other items recovered from the sea after long exposure times often suffer from graphitisation but marine archaeologists are now aware of the problem and keep the objects wet until adequate preservation is possible.

The Holland 1, however, contained little cast iron; therefore the rigorous cleaning and coating procedure was not strictly necessary. The hull of the vessel is fabricated from steel and appears to have been some 14 mm thick. The thickness of steel is still present in some areas, but other areas exhibited corrosion typically to a depth of 4-6 mm. Complete perforation has occurred in several places. The superstructure, on the other hand, is made from lighter gauge steel and wrought iron, and much of it has been lost. In addition, there are some cast iron flanges in which graphitisation had removed some 90% of the original iron, leaving rust overlying a graphite network according to the mechanism outlined above.

Several non-ferrous metals were also used in the construction of this submarine including zinc, lead and copper alloys. The torpedo tube, the conning hatch assembly and the engine exhaust ports were made of bronze or gun metal; we did not remove samples for laboratory analysis in order to minimise the damage to the vessel. The bronze fittings showed no signs of corrosion.

Let us now consider the observed corrosion rate of steel to establish whether in the Holland 1 it was, indeed, slower than would be expected from theory. The corrosion rate of carbon steels in quiet, aerated sea water (sea water is fully aerated at a depth of 200 feet, the oxygen content being some 8-10 ppm, depending on the temperature) is typically 0.125-0.250 mm/yr. In the absence of any protective paint coating it is possible that the steel hull could be corroded through in 28 years, assuming corrosion to take place from the inside as well as the outside. (Despite the fact that the hatch was apparently closed when the Holland 1 was recovered, there is evidence that the water inside had been changed many times whilst she was on the sea bed, see below.) It is clear, therefore, that the Holland 1 was not in contact with aerated sea water for much of its 69 years of immersion.

Three possible explanations are that the submarine was buried in silt, the paint coatings on the hull were extremely protective or a scale was formed on the hull. The recovery team have stated that the Holland 1 was not buried but, since the depth immersion corresponds with the normal wave base, it is quite possible that severe storms could have caused the vessel to be covered and uncovered. It is unlikely that the paints on the hull would have lasted very long, especially as it had probably not been painted for some time before it was scrapped. The reason for the relative lack of corrosion of the Holland 1 was that it was covered with scale.

The scale covered the submarine inside and out. Most of it had been removed by the cleaning procedure, but samples were obtained for laboratory analysis from regions of difficult access. This scale is buff coloured, hard, adherent and 3-6 mm in thickness. Chemical analysis indicated that it consists largely of calcium carbonate ($CaCO_3$) with some cuprous sulphide (CU_2S) present in the samples removed from the torpedo tube. Beneath the $CaCO_3$ constituent of the scale are metallic lead and some zinc salts. It is presumed that these are derived from the remnants of the paints on the hull. X-ray diffraction of the scale showed that the $CaCO_3$ is present mostly as aragonite with only a small proportion of the more stable calcite. $CaCO_3$ scales are known to be protective towards carbon steels provided they are adherent and continuous.

Generally there are four possible mechanisms for the deposition of $CaCO_3$ from sea water:

1. Chemical deposition from sea water supersaturated with $CaCO_3$. This occurs in tropical conditions and, to a lesser extent, in other warm areas such as in the Gulf Stream off the west coast of Ireland. However, it has not been reported off the English or French coasts anywhere near the Channel.

2. "Electrochemical" deposition. This takes place where local changes in pH of the sea water occur, due usually to the oxygen reduction reaction at the cathodic site when galvanic corrosion is proceeding. Most cathodically protected structures in the sea (ships hulls, oil platform jackets, etc) accumulate "calcareous" deposits, especially when overprotected (ie the potential depressed to a too negative value). Very little galvanic corrosion appears to have taken place on the Holland 1, despite the significant quantity of non-ferrous metal used in its construction; hence it is believed that this mechanism is not responsible for the formation of much of this deposit.

3. Deposition by algae and coral-like marine life. This is unlikely in the circumstances: firstly, corals are suspension feeders and would probably, therefore, produce more scale on the outside that on the inside of the hull - the scale is virtually identical in thickness and type inside and outside. Secondly, algae are unlikely to be active at the depth concerned, even though aphotic types exist which would be capable of living in the dark conditions inside the wreck.

4. Biogenic deposition of $CaCO_3$. This mechanism is believed to be responsible for the presence of the $CaCO_3$ scale on the Holland 1. Anaerobic sulphate reducing bacteria (SRB) such as desulphovibrio are very common in the Channel, inhabiting the sea-bed muds and other areas where deoxygenated conditions exist. Needing only a food (carbon) source and sulphur in a high oxidation state (typically sulphate ions), SRB thrive in conditions of limited oxygen availabilty and produce sulphide ions. Sulphide ions are very corrosive towards mild steel and are often responsible for the rapid corrosion of steel piles in estuarine mud, water injection tubing in oil wells where secondary recovery is in operation, and even tinned-steel food containers. Additionally, because of the high pH conditions produced by the sulphide ions, $CaCO_3$ is deposited locally.

For the necessary anaerobic conditions to have been created inside and outside the Holland 1, two alternatives are postulated. Firstly, that the vessel was silted on the outside, with oxygen depletion inside occurring by slight corrosion or by aerobic activity of the marine life inside. A second possibility is that the inside and outside of the vessel were fouled by marine organisms: this is often sufficient for SRB to become active between the hull and the fouling. After an unknown period of scale deposition it appears that the Holland 1 was uncovered, presumably by wave action, probably during a storm since, as mentioned above, the depth at which the vessel was found corresponds with the normal wave base and only severe storms would cause significant shifting of the sea bed. After that time the vessel was colonised by the marine life typical of the area, Devonshire cup-corals (Caryophyllia smithii), saddle oysters, bryozoa, queen scallops (Chlamys distorta), barnacles and polychaete worms (Hydroides norvegica). The scale on the outside of the vessel appeared to have 99% cover of organisms, calcareous tubes or exoskeletons, none more than ten years old, with the polychaete worms responsible for some 90% of the cover. This dense coating of tubes and shells would be expected to commence within a month or so of exposure to the sea (in the peak growing season of May to September), complete cover taking from 1-3 years.

The presence of the $CaCO_3$ scale confirms that the vessel saw deoxygenated conditions, ie it was most likely to have been silted up for much of its 69 years under the sea.

It is quite likely, therefore, that the main reason for the Holland 1 being in such good condition in parts is that silting occurred soon after sinking. The vessel was almost upright during its time on the sea bed. Deposits with the appearance of stalactites in the torpedo tube indicate a 10-15° list to starboard; hence the superstructure could easily have projected through the silt into the sea, thus accounting for what may have been more rapid corrosion. Evidence which supports the explanation outlined here for the calcium carbonate deposition is the presence of Cu_2S in the deposit in the torpedo tube (the copper originating in the copper alloy of the tube) and the strong smell of hydrogen sulphide in the Holland 1 at the time of recovery, reported to us by one of the first people to enter it in dry dock. By way of confirmation, samples of sludge from the bilges were examined in our laboratory and found to be strongly anaerobic inside, providing ideal SRB growing conditions.

There are, as yet, certain unexplained features. The total quantity of calcium carbonate estimated to line the inside of the vessel is between 1 and 2 tonnes, but the calcium content (as $CaCO_3$) of the sea water inside the vessel is only <u>ca</u> 0.1 tonne. If the calcium can be extracted from the sea water at an efficiency of 10%, some 100-200 changes of water must have occurred in the deposition phase of the submarine's 69 year immersion. It was stated by the recovery team that the conning tower hatch was closed: this leaves only a few small openings for the exchange of sea water to take place.

The significant quantities of zinc compounds and metallic lead found underneath the $CaCO_3$ scale are presumed to originate from paint coatings used on the hull, but it has not been possible as yet to check this with the paint suppliers.

It is not known whether or not the batteries in the Holland 1 contained acid at the time of sinking. If the batteries were charged with sulphuric acid, the initial pH of the water inside the submarine could have been below 4, assuming the loss of acid to be significantly faster than the exchange of sea water inside and outside the submarine. Sea water of this acidity is very corrosive and calcium carbonate deposition cannot take place.

The Holland 1, as well as being an interesting item of submarine and naval history, provided an interesting exercise in corrosion investigation. Corrosion has occurred to varying extents, the maximum calculated corrosion rate is typical of corrosion in aerated sea water, but much of the steel has exhibited significantly lower corrosion rates, though not as low as we were led to believe from reports. The factor responsible for the lower corrosion rates is the general coverage by calcium carbonate scale. This appears to have been biogenically deposited as a result of anaerobic bacterial activity. Similar mechanisms have been invoked for calcium carbonate deposition off the west coast of Australia and Islay, Scotland.[1]

There are no corrosion lessons to be learnt from this investigation but it is nice to be reminded that nature's ways of corrosion prevention may often be much more effective than those devised by man.

ACKNOWLEDGEMENTS

Thanks go to the staff of the Corrosion and Protection Centre and CAPCIS at UMIST, and to Tim Hopkins, Dave Plant and Tony Adams at the Geology Department and Steve Hawkins in the Zoology Department at Manchester University. Special thanks to the staff of the Royal Navy Submarine Museum for the opportunity of investigating the Holland 1.

REFERENCES

1. Adams, A.E. and Schofield, K., Recent submarine aragonite, magnesian calcite and haematite cements in a gravel from Islay, Scotland, J. Sedimentary Petrology, 53 (2) (1983), pp. 417-521.

VAPOUR PHASE INHIBITORS: A Review of their use for Corroded Copper and Copper Alloys

Linda G. Scheifler

Vapour phase inhibitors (VPIs) protect against metal corrosion, in the presence of moisture and oxygen, through the emission of vapours from a solid or liquid substance. Unlike other types of inhibitors, direct contact between the metal and inhibitor supply is not necessary. VPI must be used in a contained space where an equilibrium, determined by the partial vapour pressure, can be maintained so that the vapours are in sufficient concentration to provide protection. Two important features of vapour phase inhibitors make them of special interest to conservators. Corrosion protection is maintained as long as the VPI supply is sufficient; yet there is no permanent adsorption, thereby eliminating the problem of removing the protective substance from the object at a later date.

The first vapour phase inhibitors were developed during World War II in order to extend the shelf life of ferrous metal products, by the impregnation of packing materials with an inhibitor.[1,2,3] Later, they were used on a large scale with the intention of inhibiting the corrosion of industrial storage and shipping containers during periods of disuse.[4] In both cases, the focus was on the corrosion of ferrous metals, but from these early tests it was established that many non-ferrous metals actually corroded more in the presence of commonly used VPI such as cyclohexylamine carbonate and dicyclohexylamine nitrite. Volatile corrosion inhibitors which can protect a wide range of metals are now emerging after years of development.[5,6,7] However, in industry it is the effectiveness of VPI on clean metal surfaces which is tested, with the aim of extending shelf life or use period. Museums, on the other hand, are filled with corroded metal artefacts such as bronzes which are to be maintained with their patina intact. Thus on the basis of most existing research, we do not know whether the VPI will inhibit the further corrosion of corroded or patinated museum artefacts.

Although sparsely mentioned in Abstracts of Technical Studies in Art and Archaeology, 1943-52[8] and in Art and Archaeology Technical Abstracts, 1952-83,[9] vapour phase inhibitors were discussed in the conservation/museum literature as far back as 1953, when Mr. Leo Biek, then Head of the Ancient Monuments Laboratory, described in the Museums Journal[10] a vapour phase inhibitor produced by Shell. Stambolov and Moll[11] reported testing a VPI on a degreased copper plate yielding positive results. The next stage should be to test VPI on corroded samples.

This paper describes tests performed to determine the efficacy of selected vapour phase inhibitors in preventing further corrosion of copper and copper alloyed samples containing oxide, sulphate, and chloride corrosion products.

THE MECHANISM OF VAPOUR PHASE INHIBITOR PROTECTION

Vapour phase inhibitors fall into four main chemical types: amines, amine nitrites, amine carbonates, and esters of acids.[12,13] Chemicals from these categories which can be utilized must demonstrate, practically speaking, certain features.

- They must be stable under the conditions of use; that is they must be effective under tropical and/or polluted conditions, both gaseous and particulate.

- They should be sufficiently volatile for rapid diffusion, but it must be possible to maintain an adequate concentration for protection over a reasonable period of time (one to two years).

- Inhibitor toxicity must be low, especially if it is

possible for it to permeate through the container walls. Vapour concentration levels should be monitored wherever they are employed on any scale.

- Various museum applications require the availability of vapour phase inhibitors in both solid and liquid forms. The liquid form, for example, could be used to impregnate a textile in a display case where containers of crystals may be considered unsightly.

The volatility of these inhibitors transports the active agent to the metal surface. Although an increase in temperature usually encourages an increase in corrosive activity on metal surfaces, the volatilization rate of vapour phase inhibitors is also temperature dependent; more inhibitor molecules evaporate at higher temperature. Thus, VPI "have a property of self-adjustment to the degree of aggressiveness of their environment, which explains their effectiveness over a relatively wide temperature range."[6]

The method by which vapour phase inhibitors perform their function is not yet fully understood.

Autoradiographic studies[14] show that dissociated cations and anions of the vapour phase inhibitor molecule are concentrated at the cathodic and anodic areas, respectively. Beyond this elementary consensus, there are conflicting opinions as to whether dissociation occurs in the vapour phase, or within the moisture layer on the metal surface, and as to how the passivation of reactive metallic sites is achieved. Several sources[12,15,16] characterize the mechanism of VPI adsorption as a combination of chemisorption, thereby highly specific and more or less irreversibly adsorbed, and physical adsorption, thereby non-specific and reversibly adsorbed. Yet the chemisorbed layers cannot be considered as permanent protective films, since the bond between the adsorbed inhibitor and the metal surface is not strong enough to prevent the inhibitor from leaving freely, as it does, from the metal surface.[17] Thus, once the source of the vapours is depleted or removed, no protection is afforded to the metal.

In spite of much industrial testing, doubt remains as to their efficacy on corroded metal surfaces. The surface geometry of the corrosion layer, its porosity and hygroscopicity must influence the ability of VPI to passivate reactive sites at the metal surface.

CORROSION TESTS

The test was designed to examine the effectiveness of three vapour phase inhibitors in inhibiting corrosion on copper and four of its alloys. Since the objective was to simulate artefact specimens, the metal specimens were prepared with one of the following: an uncorroded surface, a copper oxide, a copper sulphate, or a copper chloride film. Thus for each of the three VPI described in the following section, twenty metal coupons were tested: five metals, each with four surface preparations. To provide controls against which weight changes could be compared, two more identical sets of twenty coupons were made.

THE VAPOUR PHASE INHIBITORS

Three vapour phase inhibitors were selected on the basis of their volatility, solubility, and availability. Safety and convenience were also taken into consideration since servicing of display cases and handling in storage areas could involve the technician and curator.

It was deemed desirable to test a proprietary product since it would be easy to obtain in a ready-to-use form. Zerust Vapour Capsules were selected. Although the complete chemical formulation was not disclosed by the manufacturer, the active VPI ingredient was identified as complex ammonium salts[18] which ionize into complex ammonium ions.[19] Zerust vapour is water soluble and claims to "reduce, not halt, corrosion for ferrous, non-ferrous, and bimetallic systems.[20] Although the manufacturer recommends the use of Zerust Vapour Capsules with copper and brass, the literature[17,21] says that amines form water soluble complexes with copper and copper alloys and are, therefore, aggressive towards them. The following restrictions are listed:[20]

- Zerust should not be used repeatedly above 82°C (180°F).
- Zerust capsules should not contact water.
- Zerust will not provide protection from corrosion caused by direct contact of the metal with water, acids, etc.
- Zerust may adversely affect the coating on photographic lenses
- Over an extended period of time, Zerust vapour will yellow paper. Do not use in proximity of antique books, valuable documents, etc.
- Do not use Zerust in close proximity of turquoise as it tends to "age" this semi-precious gem.

The information is conflicting (utilizes amines, yet protects non-ferrous metals; "ages" turquoise - a basic copper-aluminium phosphate hydrate) but Zerust was considered a candidate for testing because of its claim of effectiveness and convenience.

Secondly, n-butyl benzoate was selected.[12,22] This ester of benzoic acid is a thick, oily liquid, practically immiscible with water.[23] Although mentioned infrequently in the literature, this compound is cited as being effective because benzoate ions either furnish protection for copper and other non-ferrous metals[7,17] or form organo-metallic compounds with copper which in turn protect the metal from further corrosion.[24] Another citation suggested that benzoates do not aggravate corrosion for non-ferrous metals and in some cases appreciably reduce it.[25] While it was uncertain how much protection this compound would afford, it too was considered a suitable test material because it provides an example of a liquid, non water-soluble substance.

Finally, benzotriazole (BTA) was tested. This product has been used in conservation since 1965,[26,27] but primarily as a chemical bonded to the surface via application in solution. BTA impregnated papers, however, are also used in industry to provide a vapour phase inhibition[28,29,30] at room temperature in closed packages. Little mention is given of the mechanism by which benzotriazole works as a vapour phase inhibitor but it is assumed that BTA inhibits corrosion by chemisorption and physical adsorption over short periods of time at room temperature.[29]

Fig. 1 summarizes the properties and structures of the inhibitors selected for testing.

VPI	Name	Structure/Formula	Solubility	BP/MP/VP
1	Zerust© Vapour Capsules	Complex ammonium salts Acid NH$_4$	water	VP 10^{-4} mmHg at 20°C
2	n-butyl benzoate	-COO-CH_2-CH_2-CH_2-CH_3	alcohol ether	BP 248°C
3	benzotriazole	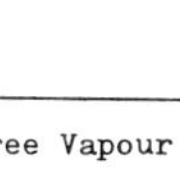	alcohol benzene toluene chloroform DMF water, sparingly	MP 98.5°C

Fig. 1 The Three Vapour Phase Inhibitors Tested

THE METAL TEST COUPONS

Copper and four copper alloys were selected on the basis of availability as well as variety. Pure copper, one binary, one tertiary, and two quaternary alloys were tested: a brass, a bronze and variants of both. The intention was to monitor the effects of the inhibitor on the amphoteric constituents of the alloy, in case destannification or dezincification occurred. Their compositions are summarized in Fig. 2.

Metal	%Cu	%Sn	%Zn	%Pb	%P
1	92.315	7.60		.005	.08
2	80	10		10	
3	100				
4	85	5	5	5	
5	70		30		

Fig 2 The Metals Tested: weight percents of the metal composition

The American Society for Testing Materials (ASTM) Standard Recommended Practice for Preparing, Cleaning, and Evaluating Corrosion Test Specimens,[31] Laboratory Immersion Corrosion Testing of Metals,[32] and Conducting Atmospheric Corrosion Tests on Metals[33] were adopted in part for sample preparation. When corrosion growth is to be detected by changes in weight, a large surface-to-mass ratio with a small edge area/total area ratio is desirable. Ideally, a circular specimen of minimum thickness is used. However, due to the constraints of the available equipment, circular specimens could not be produced and strip coupons were fabricated. As suggested in ASTM Standard G31-72,[32] coupons with two square inches (12.9 cm^2) of surface area and one-sixteenth inch (.16 cm) thick were produced by rolling to the desired thickness. There are several advantages to rolling rather than machine cutting, such as minimal variation in corrosion resistance of the surface and regulation of any desired thickness.[34] A final annealing step returned the crystalline structure of the metal to that of cast metal artefacts.

It was decided that interpretation of the results would be simplified if metals of known composition had one corrosion product per sample rather than multiple layers. In order to keep track of the various metals and corrosion products, a series of notches were cut, by which any one coupon could be identified.

All coupons were degreased in acetone and cleaned in a 10% by volume solution of hydrochloric acid (HCl, sp. gr. 1.84) in deionized water. After rinsing in acetone, the samples were placed in a desiccator containing dried silica gel (approximately 20% relative humidity) for twenty-four hours and then weighed. All samples were annealed by placing them in a furnace at 600°C for four hours, at which time the furnace was turned off and the samples left inside for fourteen hours to cool. The samples were then removed (the furnace had cooled down to 250°C) and air cooled for five hours. The oxide layers, in various shades of grey, were thus formed. Scanning electron microscopy studies later showed that these oxides were thin and unprotective and that dezincification had occurred on Metal 5 coupon. X-ray diffraction studies, summarized in Fig. 3, show both copper oxides as well as the oxides of alloying constituents.

The uncorroded, sulphate, and chloride coupons were cleaned of oxide layers formed during annealing by immersing the samples in 10% by volume solution of HCl for three to four hours. Upon removal from the acid, the coupons were scrubbed clean with a toothbrush in distilled water and immediately dried through acetone. The uncorroded set of coupons was thus ready for testing. As a result of this acid cleaning there were small reddish patches on all brass (Metal 5) coupons. While conforming to the ASTM Standards given at the beginning of this section, these reddish patches were probably due to the redeposition of copper onto the surface of the coupons when the concentration of copper in the acid bath reached a certain level or possibly due to a small amount of dezincification on the surface.

In order to produce sulphate and chloride corrosion

Metal	Oxide Film	Sulphate Film	Chloride Film
1	CuO Cu_2O SnO	not identifiable	CuCl $CuCl_2$ H_2O $CuCl_2$ 3 $Cu(OH)_2$
2	CuO PbO ? SnO_2 ?	not identifiable	CuCl $CuCl_2$ $2H_2O$ $CuCl_2$ 3 $Cu(OH)_2$
3	CuO Cu_2O	not identifiable	$Cu_2(OH)_3Cl$ $CuCl_2$ $2H_2O$ $CuCl_2$ 3 $Cu(OH)_2$
4	CuO CuO_2	not identifiable	CuCl $CuCl_2$ H_2O $CuCl_2$ 3 $Cu(OH)_2$ $SnCl_2$
5	ZnO	not identifiable	$CuCl_2$ 3 $Cu(OH)_2$

Fig 3 The Surface Preparation: X-ray Diffraction Analysis of Compounds Formed

products, the samples were placed in sealed containers in the presence of HCl and H_2SO_3 vapours for 21 days. On the twenty-first day after their exposure the samples were removed from the containers and placed in a desiccator. Subsequent scanning electron microscopy (SEM) and X-ray diffraction (see Fig. 3) studies showed that a variety of cuprous and cupric chlorides were formed in anhydrous, hydrous and basic forms and that destannification had occurred on Metal 4. Sulphate X-ray patterns were very confusing and could not be identified, although corrosion products had been formed. In SEM studies at 500x and 1000x magnification a very thin incoherent layer could be seen on the sulphate sample from Metal 4.

All the prepared coupons were now ready for testing and were placed in a desiccator for two days.

THE TEST PROCEDURES

The effectiveness of the vapour phase inhibitors was measured by monitoring the weight of each coupon before and after testing and comparing the weight changes with those of the control coupons. To ensure that all changes in weight would be recorded, each coupon was placed on a glass microscope slide. To minimize weight changes in the glass slides, they were placed in a desiccator for twenty-four hours before the initial weights were recorded.

In review, each of the five identical test sets were composed of twenty metal coupons: five different types of copper metals (see Fig. 2) each with one of four surface preparations - uncorroded, oxide, sulphate and chloride. The test conditions for each of the five sets will now be outlined.

Two control groups compared the effects of sulphur-contaminated and non-contaminated environments. For Control A the coupons were placed on a glass shelf in a glass baking dish with 150 ml of water in the bottom. Control B coupons included, in addition, a 25 ml dish of H_2SO_4 to create a mildly aggressive, sulphur-containing environment. Because the three vapour phase inhibitor test sets would also contain sulphuric acid, Control B will be used for comparing the test data. Control A is useful in comparing the rate of corrosion in environments with and without sulphur in order to measure how much more aggressive the sulphur-containing environment was in this test.

The coupons designated for testing the Zerust Vapour Capsules were similarly placed in another baking dish which contained 150 ml of water at the bottom, a 25 ml dish of H_2SO_4, and the Zerust capsule. The conditions for the other two inhibitors, n-butyl benzoate and benzotriazole, were identical to that of Zerust.

All the glass dishes were covered and sealed with Dow Silicone Stopcock Grease, secured with masking tape and placed in an oven at 40°C. This temperature was selected because it is near the top of the temperature range to which most artefacts would be subjected, especially during transit. A more aggressive environment could have been created if cyclic temperature fluctuations occurred. It was felt, however, that this would also have been too severe a test for the materials. The atmosphere within the containers was at the saturation point, that is near 100% relative humidity at 40°C, since there was always water at the bottom of the container.

In order to monitor the rate of corrosion, Controls A and B were removed from the oven for one hour on the fifth and fifteenth days and weighed (without being placed in the desiccator for twenty-four hours) in order to get an approximate idea of the weight change. The testing was stopped on the twentieth day since the monitoring of Controls A and B indicated significant weight changes and since preliminary tests carried out suggested a twenty-day test period. The samples were finally weighed after being in a desiccator for twenty-four hours. All weights are considered significant to ± 0.00005 g.

RESULTS AND DISCUSSION

The results of the test are summarized in Fig. 4 while Fig. 5 illustrates the main findings. Statistical tests were not applied to the measurements but significant visual changes were observed. Since active corrosion on museum objects is routinely detected (and considered significant) by visual inspection rather than by weight measurements, the numbers are used to suggest the degree of corrosion observed. Thus, weight changes could be considered significant since they correspond to significant visual changes.

Note should be made that the Zerust - Metal 4 sulphate and Metal 5 sulphate coupons and the n-butyl benzoate - Metal 1 uncorroded coupon were contaminated during the experiment by water contact. Yet only Zerust - Metal 4 sulphate coupon had a net weight loss by the end of the experiment. It is unclear why this was the case but perhaps the corrosion rates of Metal 5 and Metal 1 were sufficiently high to register an overall weight gain in spite of the fact that some corrosion products had been dissolved into the water.

Coupon	Control A	Control B	Zerust	n-butyl benzoate	BTA
Metal 1					
uncorroded	0.11	0.02	0.07	0.05	0.03
oxide	- 0.06	0.00	0.01	0.02	0.00
sulphate	0.04	- 1.25	0.05	0.03	0.03
chloride	1.73	0.94	2.04	2.97	2.35
Metal 2					
uncorroded	0.00	0.02	0.12	0.02	0.00
oxide	0.03	- 0.27	0.04	0.02	0.03
sulphate	0.08	0.03	0.08	0.04	0.04
chloride	1.06	0.83	1.12	1.75	1.97
Metal 3					
uncorroded	0.03	0.00	0.01	0.00	0.01
oxide	0.00	- 0.02	- 0.02	0.16	0.00
sulphate	0.01	0.00	5.37	0.00	0.01
chloride	0.86	0.97	2.09	1.58	3.65
Metal 4					
uncorroded	0.01	0.04	0.04	0.01	0.03
oxide	0.02	0.01	0.04	0.00	0.01
sulphate	0.03	0.17	- 4.44	0.02	0.03
chloride	1.20	1.70	1.97	1.60	1.34
Metal 5					
uncorroded	0.01	0.02	0.06	0.00	0.01
oxide	0.00	0.00	0.04	0.00	0.00
sulphate	0.01	0.02	0.08	0.05	0.03
chloride	0.83	0.90	0.85	0.85	1.15

Fig 4 Percent Weight Gain of Each Coupon

Some trends or patterns can be seen in the figures given in Fig. 4. The uncorroded and oxide coupons show little corrosion in the course of twenty days. However, for the sulphate film on Metal 3, corrosion enhancement is found in the presence of Zerust. For the chloride coupons it appears that the control groups as well as all three vapour phase inhibitor protected samples underwent corrosion during the twenty days. Control A and Control B show that the uncorroded, oxide, and sulphate films are relatively inactive as compared with the chloride films. Zerust seems satisfactory for the uncorroded and oxide film, unpredictable for the sulphate, and ineffectual on the chloride film. n-Butyl benzoate and benzotriazole seem adequate for all but the chloride samples.

By comparing the percent weight change of the vapour phase inhibitor coupons to that of Control B coupons, it was noticed that the uncorroded, oxide, and sulphate coupons corroded at the same rate as their Control B counterparts, with the previously noted exception of the corrosion enhancement of one sulphate coupon. However, for the chloride coupons a new pattern emerges. Metals 1, 2, and 3 underwent enhanced corrosion in the presence of all three VPI while Metals 4 and 5 corroded less than their Control B counterparts, although corrosion occurred.

The last figure (Fig. 5) summarizes the overall capabilities of each VPI. Zerust seems to enhance the corrosion rate of the Metal 3 sulphate coupons as well as the chloride coupons of Metals 1, 2, 3 and 4. Only the Metal 5 chloride coupons corroded less than their Control B counterparts. n-Butyl benzoate and benzotriazole did not corrode, oxide and sulphate coupons corroded about the same

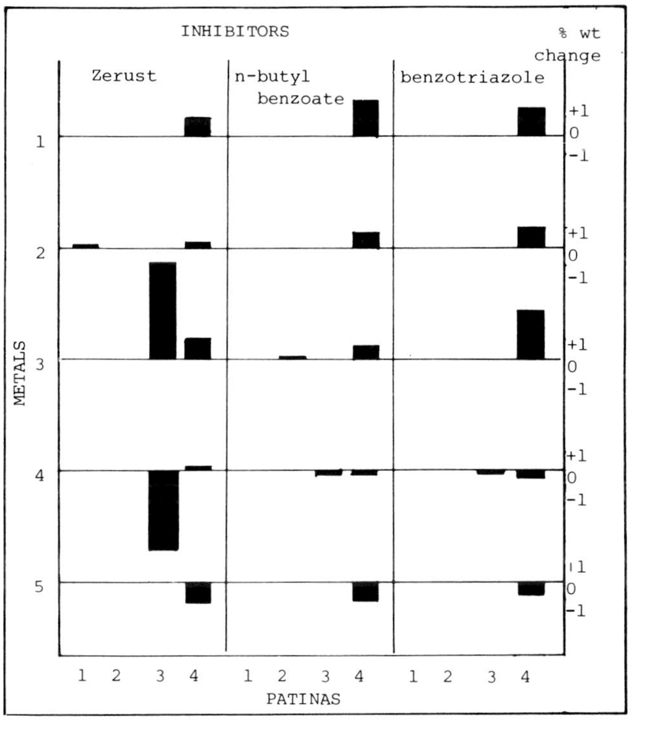

1=uncorroded 2=oxide 3=sulphate 4=chloride

Fig 5 Weight change of inhibited coupons minus weight change of Control B, after 20 day test.

A negative change indicates that the inhibited sample corroded less than did the control B sample; a positive change that corrosion was more.

as their Control B counterparts. However, for both vapour phase inhibitors the chloride coupons for Metals 1, 2 and 3 underwent corrosion enhancement while Metals 4 and 5 corroded less than their Control B counterparts.

CONCLUSIONS AND RECOMMENDATIONS

The performances of the three vapour phase inhibitors with uncorroded coupons and those with oxide or sulphate films showed no significant variations. Yet a longer test period or a more sensitive method of corrosion detection would also allow a more conclusive evaluation of VPI effectiveness upon these surface preparations.

The chloride samples, however, yielded a different pattern. Their protection seems to be alloy specific, with Metals 4 and 5 somewhat protected while the corrosion of Metals 1, 2 and 3 was somewhat stimulated. Although Metals 4 and 5 were somewhat protected as compared to Control B, the rate of corrosion as shown by the percent weight gain (Fig. 5) reveals that all chloride samples underwent measurable corrosion during the twenty-day experiment. Therefore, while the VPI did afford some protection, it was not considered enough to arrest the alarmingly rapid rate of corrosion for chloride contaminated coupons. Since the inhibition of the chloride samples is of concern to conservators, it does not seem safe to use VPI with chloride contaminated materials since it either enhances the corrosion rate or, at best, provides a reduced yet unacceptable rate of corrosion.

Therefore, it is recommended that other methods of corrosion prevention be employed in favour of the vapour phase inhibitors tested. Control of environmental factors such as humidity, volatile contaminants and particulate matter can be achieved with silica gel, surface coatings, vapour absorbing papers, and clean storage conditions.[35] Whenever vapour phase inhibitors are utilized, however, note should be made that silica gel not only absorbs water vapour but other vapours, including that of VPI. Therefore do not use vapour phase inhibitors with a silica gel back-up.[4]

I wish to conclude by encouraging others to conduct corrosion tests relevant to conservation applications. It is hoped that improvements will be made upon these procedures and reported to the profession. As a result of conducting these tests, I discovered the lack of a sensitive corrosion detection system which is both inexpensive and convenient to use for corroded test coupons. Upon development of such a technique, testing of conservation treatments and materials would be easier. Secondly, Conservation Testing Standards should ideally be submitted to ASTM. This would permit replication and comparison of test results similar to that which exists for industrial testing. The ASTM guidelines were followed in this test, yet I would now recommend that acids be avoided as pickling agents in the sample preparation of copper and copper alloy conservation corrosion studies. Undesirable surface films could be removed instead by abrasion with papers and polishing powders, thereby limiting the possibility of dealloying the samples.

APPENDIX

A survey of the vapour phase inhibitors discussed in the literature for copper and copper alloys.

Acetamide

does not inhibit[36]

Allyl Thiourea

does inhibit[37]

Ammonium Benzoate

does inhibit[38,39]

Ammonium Carbonate

does inhibit[12]

Benzamide Dimethylformamide

does not inhibit[36]

Benzotriazole

does inhibit[28,29,30,36,40]

n-Butyl Benzoate

does inhibit[17,41]

Cyclohexylamine Carbonate (CHC)

attacks[1,2,13,35,42]
does inhibit (if in contact with metal)[43]
does inhibit (if no water condensation occurs)[44]

Dibutyl Adipate

does inhibit (brass)[38]

Dibutyl Phthalate

partially inhibits[45]

Dicyclohexylamine Carbonate

does inhibit[46]

Dicyclohexylamine (Nitrite (DICHAN)

does inhibit[36]
attacks[3,4,10,12,38,39,46,47,48]

Diisopropyl Ammonium Nitrite

attacks[12]
does inhibit (if in a specific mixture)[46]

m-Dinitrobenzene

does inhibit[7]
neither inhibits, nor attacks[49]

Esters of Isoamyl and Ethyl Chromic Acid

does inhibit[50,51]
does not inhibit[5]

Ethanolamine Carbonate
Ethanolamine Benzoate

attacks[46]

Ethylene Diamine

does inhibit[36]

Hexamethylenemine 3,5 dinitrobenzoate

does inhibit[7]

Hexamethylenemine o-nitrobenzoate
Hexamethylenemine m-nitrobenzoate

does inhibit[7]

Lauryl Alcohol

partially inhibits[45]

Methylmorpholine

does inhibit[36]

Monoethanolamine

does inhibit[36]

Monoethanolamine Carbonate

neither inhibits nor attacks[36]
attacks[12]

Morpholine

does inhibit[36]

β-Napthol

neither inhibits, nor attacks[49]

p-Nitrophenol
m-Nitrophenol

attacks[49]

Phenol

does inhibit[36]

Phenyl Thiourea

does inhibit[37]

β-Picoline

does inhibit[36]

Piperdine 3,5 dinitrobenzoale

does inhibit[7,36,38]

Propylene Diamine

does inhibit[36]

Pyridine

does inhibit[36]

Quinoline

does inhibit[36]

Sodium Adipate

does inhibit[38]

Tetramethyl Thiouram Monosulfide

does inhibit[37]

SAFETY

Zerust Vapour Capsules

according to the Materials Safety Data Sheet[18] filed by Northern Instruments Corporation to the US Department of Labor, Occupational Safety and Health Administration:

"Health Hazard Data"

Threshold Limit Value: Not established - chemicals moderately toxic, vapours non-toxic

"Special Protection Information"

Respiratory Protection - dust mask
Ventilation - adequate local exhaust
Protective Gloves - rubber
Eye Protection - safety glasses

(NB The Materials Safety Data Sheet is available directly from Northern Instruments Corporation. See Materials and Suppliers section)

n-Butyl Benzoate (benzoic acid butyl ester)

Lethal dose: a dose which is lethal to 50% of rats tested orally: 5.14g/kg.[23]
Avoid skin contact and inhalation.

Benzotriazole

Avoid skin contact and inhalation.
See conservation literature.

ACKNOWLEDGEMENTS

The testing described here was undertaken while the author was a conservation student at Queen's University, Kingston, Ontario, Canada. I would like to thank the faculty of the Art Conservation Programme and Department of Metallurgy for providing encouragement and assistance throughout the project. X-ray diffraction and scanning electron microscopy analysis was done under the guidance of Dr. Heyding of the Chemistry Department and Mr. David Kempson of the Geology Department. Special thanks are given to Dr. Ursula Franklin, University of Toronto, for the valuable advice given at the crucial beginning stages whereby test parameters were established.

MATERIALS AND SUPPLIERS

Zerust Vapour Capsules: Northern Instruments Corporation, 6680 North Hodgson Road, Lino Lakes, Minnesota 55015, USA; telephone: (612) 784-1250.

n-Butyl Benzoate ($C_{11}H_{14}O_2$): chemical supply houses.

Benzotriazole ($C_6H_5N_3$): chemical supply houses.

Metal 1: Bronze (weight percents: 92.315 Cu, 7.60 Sn, .005 Pb, .08 P), Mr. P A Scully, Noranda Metal Industries Limited, Postal Box 1158, Station A, Montreal, Quebec H3C 2Y4, Canada; telephone: (514) 871-9650.

Metal 4: Bronze/Brass (weight percents: 85 Cu, 5 Sn, 5 Zn, 5 Pb), Metals and Alloys Limited, 1611 Bercy Street, Montreal, Quebec, Canada; telephone: (514) 526-2561.

Metals 2, 3, and 5: not available commercially. These metals were formed by mixing pure metals by weight percents.

REFERENCES

1. Bennister, H.L., Vapour phase inhibitors, Research (London): science and its application in industry, 5 (1952), pp. 424-432.

2. Stroud, E.G. and Vernon, W.H.J., The prevention of corrosion in packaging: III. Vapour-phase inhibitors, Journal of Applied Chemistry, 2 (1952), pp. 178-184.

3. Wachter, A., Skei, T. and Stillman, N., Dicyclohexylammonium nitrite, a volatile inhibitor for corrosion preventative packaging, Corrosion, 7 (1951), pp. 284-294.

4. Sexton, S.E., Vapour-phase inhibition of atmospheric corrosion, Corrosion Technology, 7 (1960), pp. 147-150, 162.

5. Levin, S.Z., Gintzberg, S.A., Dinner, I.S. and Kuchinsky, V.N., Synthesis and protective action of some inhibitors on the basis of cyclohexylamine and dicyclohexylamine, Proceedings of the second European symposium on corrosion inhibitors, 22-24 September 1965, Ferrara, Italy, 2 (1966), pp. 765-776.

6. Miksic, B.A., Some aspects of metal protection by vapour phase inhibitors, Anti-corrosion methods and materials, 22 (1975), pp. 5-6.

7. Rozenfel'd, I.L. and Persiantseva, V.P., The use of inhibitors for the protection of metals from atmospheric corrosion, Protection of Metals, 2 (1966), pp. 13-17.

8. Gettens, R.J. and Usilton, B.M., Abstracts of technical studies in art and archaeology, 1943-52, Washington (1955), 553, 555, 556.

9. International Institute for Conservation, Art and Archaeology Technical Abstracts, New York (1952-83), 1-4, 3-2319, 3-2600, 3-2922, 4-3543, 4-3968, 5-4778, 5-4890, 12-1200, 13-584, 14-181, 14-541, 14(Supp)-278.

10. Biek, L. Vapour phase inhibitors (VPI), Museums Journal, 53 (1953), p. 110.

11. Stambolov, T. and Moll, E., Reduction of tarnishing of silver and copper with vapour phase inhibitors, ICOM Report 69/37, Amsterdam (1969). 4 pp.

12. Desai, M.N., Thanki, G.H. and Gandhi, M.H., Vapour phase corrosion inhibitors, a review, Australasian Corrosion Engineering, 13 (1969), pp. 19-21.

13. Rowden, S., Vapour-phase corrosion inhibitors, Corrosion Technology, 5 (1958), pp. 117-120.

14. Henriksen, J.F., Use of autoradiography for studying the mechanism of VPIs, presented to the International corrosion forum devoted exclusively to the protection and performance of materials, 22-26 March 1976, Houston, Texas. National Association of Corrosion Engineers Paper Number 85, p. 5.

15. Baker, H.R. and Zisman, W.A., Liquid and vapour corrosion inhibitors, Lubrication Engineering, 7 (1951), p. 122.

16. Hackerman, N., Recent advances in understanding of organic inhibitors, Corrosion, 18 (1962), p. 332t.

17. Rozenfel'd, I.L., Persiantseva, V.P., and Terent'ev, P.B., Mechanism of metal protection by volatile inhibitors, Proceedings of the Second International Congress on Metallic Corrosion, 1963, New York. Texas (1966), pp. 653-655.

18. U.S. Department of Labor, Occupational safety and health administration, Material Safety Data Sheet for NI-22790, (n.d.), p. 1.

19 Young, Diane, Marketing Services, Northern Instruments Corporation, personal correspondence, 4 May 1979.

20. Northern Instruments Corporation, Zerust Vapour capsules for corrosion control application guide, Lino Lakes, Minnesota (1979), pp. 1, 7, and 8.

21. Encyclopedia of Chemical Technology, 2nd ed., New York (1965), Volume 6: Corrosion Inhibitors, p. 343.

22. Vernon, W.H.J., Chemical research and corrosion control: some recent contributions of a corrosion research group, Journal of the Society of Chemical Industry, 66 (1947), p. 139.

23. Merck & Company, The Merck Index, 9th ed., Rahway, New Jersey (1976), number 1536.

24. Rozenfel'd, I.L., Enikeev, E.Kh., and Zvesdinski, K.V., Kinetics of interaction of volatile inhibitors with iron and copper surfaces, Protection of Metals, 10 (1974), p. 412.

25. Stroud, E.G. and Vernon, W.H.J., The Prevention of Corrosion in Packaging: I. The use of sodium benzoate as a corrosion-inhibitor in wrappings, Journal of Applied Chemistry, 2 (1952), p. 170.

26. Neznamova, T.G., Dobrovol'skaya, V.P., and Barranik, V.P., Compounds of the azole group as corrosion inhibitors of ferrous metals and copper. I. protective action of benzotriazole in neutral and acid solutions, Zhurnal Prikladnoi Khimi (In Russian), 38 (1965), pp. 2388-2390. (see Art and Archaeology Technical Abstracts 6-H2-17).

27. Madsen, H.B., A preliminary note on the use of benzotriazole for stabilizing bronze objects, Studies in Conservation, 12 (1967), pp. 163-167.

28. Copper Development Association, Inc., Benzotriazole: an effective corrosion inhibitor for copper alloys, Application Data Sheet, Number 108/9 (n.d.), pp. 1-4.

29. Cotton, J.B. and Scholes, I.R., Benzotriazole and related compounds as corrosion inhibitors for copper, British Corrosion Journal, 2 (1967), pp. 3-4.

30. Hallowes, A.P.C., Brighter Outlook: papers and tapes preserve surface quality of copper and its alloys, Copper, 15 (1962), p. 5.

31. American Society for Testing Materials, Standard recommended practice for preparing, cleaning and evaluating corrosion test specimens, ASTM Book of Standards, Part 10, Number G1-72, pp. 632-636.

32. American Society for Testing Materials, Standard recommended practice for laboratory immersion corrosion testing of metals, ASTM Book of Standards, Part 3, Number G31-72, pp. 882-891.

33. American Society for Testing Materials, Standard recommended practice for conducting atmospheric corrosion tests on metals, ASTM Book of Standards, Part 10, Number G50-76, pp. 878-885.

34. Champion, F.A., Corrosion Testing Procedures, 2nd ed., New York (1965), p. 10.

35. Evans, E.L. and Stroud, E.G., Prevention of corrosion in packaging and storage - control of atmospheric corrosion of metals in storage and transit, Chemistry and Industry, 9 (1957), pp. 242-248.

36. Desai, M.N., Rana, S.S., and Gandhi, M.H., Corrosion inhibitors for copper, Anti-Corrosion Methods and Materials, 18 (1971), p. 22.

37. Agarwala, V.S. and Tripathi, K.C., Some vapour phase inhibitor experiments, Materials Protection, 5 (1966), p. 26.

38. Desai, M.N., Shah, Y.C., and Gandhi, M.H., Inhibition of corrosion of brass - a review, Australasian Corrosion Engineering, 12 (1968), p. 6.

39. Miksic, B.A., Structured Questions, in Corrosion and metal artifacts - a dialogue between conservat rs and achaeologists and corrosion scientists, edited by B.F. Brown, et. al., National Bureau of Standards, Washington, D.C. (1977), p. 223.

40. Cotton, J.B., Improvements in or relating to the inhibition of corrosion of copper and copper containing

alloys, U.S. Patent Number 907,794 (1960), 3 pp.

41. Vernon, W.H.J., Chemical research and corrosion control: some recent contributions of a corrosion research group, Journal of the Society of Chemical Industry, 66 (1947), p. 139.

42. Bhowmik, S.K., A study of the preservation of metal objects by vapour phase inhibitors, Conservation in the Tropics, ed. by O.P. Agrawal, New Delhi (1974), p. 118.

43. Paul, W., Systematic experiments designed to test the corrosion-preventing action of VPI on commonly used metals, Werkstoff und Korrosion, 7 (1956), pp. 189-198.

44. Rhodes, C.A., Cyclohexylamine carbonate as a volatile corrosion inhibitor, Corrosion Prevention and Control (1957), p. 38.

45. Volrabova, E., New methods of analysis of the protective action of volatile inhibitors, Proceedings of the Third International Congress on Metallic Corrosion, Moscow, 2 (1969), p. 150.

46. Putilova, I.N., Balezin, S.A., and Barannik, V.P., Metallic Corrosion Inhibitors, New York (1960), pp. 159-162.

47. Baker, H.R., Volatile rust inhibitors, Industrial and Engineering Chemistry, 46 (1954), p. 2597.

48. Barton, K.J., The protection of cleaned iron objects, Museums Journal, 60 (1961), p. 288.

49. Rajagopalan, K.S. and Ramaseshan, G., Studies on vapour phase inhibition: Part 1 - performance of m-Dinitrobenzene and β-Naphthanol at different humidities, Journal of Scientific and Industrial Research, 19A (1960), pp. 277-279.

50. Gintsberg, S.A. and Shreider, A.V., Amine chromates and chromate esters as inhibitors of corrosion, Journal of Applied Chemistry of the USSR, 33 (1960), p. 1580.

51. Knaack, D.F. and Brooks, D., Inhibitors for temporary protection, in Corrosion Inhibitors, ed. by C.C. Nathan, Houston, Texas (1973), p. 225.

List of Contributors

J. Argo,
89 rue du Dome,
Hull,
Quebec,
Canada.

S. M. Bradley,
Research Laboratory,
British Museum,
London WC1B 3DG

H. Brinch Madsen,
det Kongelige Danske Kunstakademi,
Konservatorskolen,
Esplanaden 34,
1263 København K,
Denmark.

R. E. Child,
Welsh Folk Museum,
St Fagan's,
Cardiff.

L. G. Scheiffler,
2338 44 Avenue,
San Francisco,
California 94116,
U.S.A.

M. J. Schofield &
A. N. Rothwell,
UMIST/CAPCIS,
P.O. Box 88,
Sackville Street,
Manchester M60 1QO

Brian S. Skerry,
UMIST/CAPCIS,
P.O. Box 88,
Sackville Street,
Manchester M60 1QO

James A. Spriggs,
Conservation Laboratory,
York Archaeological Trust,
Galmanhoe Lane,
Marygate,
York YO3 7DZ

S. Turgoose,
UMIST/CAPCIS,
P.O. Box 88,
Sackville Street,
Manchester M60 1QO

Sylvia J. Turner,
Norfolk Museums Service,
Castle Museum,
Norwich NR1 3JU

P. T. Wilthew,
Ancient Monuments Laboratory,
Fortress House,
Saville Row,
London W1X 2HE